REAL
CITIES

REAL CITIES

Modernity, Space and the Phantasmagorias of City Life

STEVE PILE

London • Thousand Oaks • New Delhi

© Steve Pile 2005

First published 2005

Apart from any fair dealing for the purposes of research or private study, or criticism or review, as permitted under the Copyright, Designs and Patents Act, 1988, this publication may be reproduced, stored or transmitted in any form, or by any means, only with the prior permission in writing of the publishers, or in the case of reprographic reproduction, in accordance with the terms of licences issued by the Copyright Licensing Agency. Inquiries concerning reproduction outside those terms should be sent to the publishers.

SAGE Publications Ltd
1 Oliver's Yard
55 City Road
London EC1Y 1SP

SAGE Publications Inc.
2455 Teller Road
Thousand Oaks, California 91320

SAGE Publications India Pvt Ltd
B-42, Panchsheel Enclave
Post Box 4109
New Delhi 110 017

British Library Cataloguing in Publication data

A catalogue record for this book is available from the British Library

ISBN 0 7619 7041 X
ISBN 0 7619 7042 8 (pbk)

Library of Congress Control Number available 2004099523

Typeset by C&M Digitals (P) Ltd., Chennai, India
Printed in India at Gopsons Paper Ltd, Noida

For Tommy

Contents

Acknowledgements   x

Introduction   1
in which we explore how to explore the phantasmagorias of city life

- Introduction: what is real about cities?   1
- Psychogeographies of London: Sinclair and Keiller travel in space and time   4
- Psychogeography and the Real City   12
- The City and States of Mind   15
- Phantasmagoria and the Realness of City Life   19

1 The Dreaming City   25
in which cities turn into dreams and dreams turn into cities

1.1 Introduction   25
1.2 A Phantasmagoria of Dreams?   28
1.3 Cities of the Imagination   35
1.4 Dream Analysis and Dream-Work   41
1.5 Dream-Work and the Work of City Space   47
1.6 City Space and the Web of Dreams: Paris and London   50
1.7 Cities are Lived as Dreams: Walter Benjamin on modernity and awakening   53
1.8 Conclusion: phantasmagorias, dream-work and the fulfilment of wishes   56

2 The Magic City   59
in which we are careful about what we wish for

2.1	Introduction: the magic of city life	59
2.2	Civilisation and its Discontents: the science of the city versus the powers of magic	61
2.3	New York's Magic: super-natures, spirits and Vodou	66
2.4	Syncretic Cities: New Orleans and the Voodoo Atlantic	73
2.5	Stick it to Them: cities, corporations and Voodoo economics	84
2.6	Singapore and the Fine Art of Placement: building fortunes in the city	87
2.7	London and the Global Trade in Magic: the high cost of healing	91
2.8	Conclusion: cities, careful wishes and the occult	94

3 The Vampiric City   97
in which blood flows free

3.1	Introduction: the life-blood of the city	97
3.2	A Plague upon the West: vampire and Empire	102
3.3	London, Dracula and the edges of Empire: circulation, blood, technology	106
3.4	Mortal Cities: Dracula, blood and money	109
3.5	Singapore, Red Lips and Sharp White Teeth: sex and death	112
3.6	To be a Vampire, forever: New Orleans, blood and belonging	117
3.7	New Orleans and Vampire Tours: selling horror and faith in the vampire	123
3.8	Conclusion: blood, life and the undead circulations of city life	128

4 The Ghostly City   131
in which the city is haunted and haunting

4.1	Introduction: the haunting quality of city life	131
4.2	Remembrance of Times Past: memories, phantasms and Benjamin's Berlin	137
4.3	Seeing Dead People: childhood, the uncanny and living with loss	140
4.4	Building Sites: sighting ghosts in New Orleans, London, Singapore	143
4.5	New York Dead and Buried: the city and grief-work	148
4.6	Spooks in London: melancholia and anti-colonial struggle	151
4.7	Nightmares of the Living: ghosts versus anti-capitalist protesters in London	156
4.8	Conclusion: bury the dead, not the truth	162

Conclusion
the real dream of city life — **165**

- Introduction — 165
- The Work of Modern City Life — 166
- Occult Spatialities and City Life: New York, New Orleans, London and Singapore — 172
- Dreaming New Futures for the City — 176
- The Real of Cities and the Life of Phantasmagoria — 180

Notes — **183**

References — **197**

Index — **211**

Acknowledgements

While preparing this book, I gave many papers on dreams and ghosts, and some on vampires and magic. I would like to thank all those who contributed to the discussion of these papers. In particular, these included sessions at the 1998, 2001 and 2003 annual conferences of the Association of American Geographers and at the 2002 annual conference of the Royal Geographical Society; in the Geography Departments at Bristol, Kentucky, Louisiana State University, Sheffield, Lund, Dartmouth and Singapore; and in interdisciplinary conferences held in London (organised by Gary Bridge and Sophie Watson), Perth (organised by Jean Hillier), Minnesota (organised by Karen Till and the Place/Space research group), Birmingham ('Modernist Transactions') and Singapore (organised by John Phillips, Ryan Bishop and Wei Wei Yeo). I have benefited from some sharp questioning and insightful commentary. As a result of these papers, many have sent me references to follow up as well as more ghost (etc.) stories. If only I had the many lives necessary to take everything into account!

I have published a few pieces that were intended to be experiments in the content and style that would feed into *Real Cities*. Although substantially revised, some sections from these pieces provide the basis for parts of the book. Material has been taken from these sources: 'Sleepwalking in the Modern City: Walter Benjamin and Sigmund Freud in the world of dreams', in Gary Bridge and Sophie Watson (eds), *A Companion to the City*. 2000, Oxford: Basil Blackwell, pp. 75–86; '"The Problem of London", or, how to explore the moods of the city', in Neil Leach (ed.), *The Hieroglyphics of Space: reading and experiencing the western metropolis*. 2002, London: Routledge, pp. 203–216; 'Spectral Cities: where the repressed returns and other short stories', in Jean Hillier and Emma Rooksby (eds), *Habitus: a sense of place*. 2002, Sydney: Ashgate, pp. 219–239; 'Perpetual Returns: vampires and the ever colonised city', in Ryan Bishop, John Phillips and Wei Wei Yeo (eds), *Postcolonial Urbanism: Southeast Asian cities and global processes*. 2003, New York: Routledge, pp. 265–286; and, 'Ghosts and the City of Hope', in Loretta Lees (ed.), *The Emancipatory City?: paradoxes and possibilities*. 2004, London: Sage, pp. 210–228.

It would not have been possible to gather the range of stories that I have without the enthusiasm and support of people in the cities I have visited. In New Orleans, I was helped primarily by Dydia DeLyser, but also by Helen Regis and Rebecca Sheehan; in Singapore, by Lisa Law and by Henry Yeung and Robbie Goh; and, in Johannesburg, by Teresa Dirsuweit. I have also been fortunate to be part of a wider set of intellectual discussions – I hope people will forgive me if I do not list everyone I have talked to about this project.

I have had financial support, throughout this project from the Geography Department at the Open University. The Department also has a vibrant and supportive intellectual culture, in which I have been able to try out and develop ideas. In particular, John Allen – both as Head of Department and as a colleague – has been enormously helpful. Michèle Marsh helped to produce and collate various drafts of the book. Jan Smith, with Susie Hooley, Neeru Thakrar and Rebecca White, never tired in scanning images for me. Moreover, Jan Smith arranged the permissions for those images that required it. I am grateful to the permission holders for allowing me to reprint certain images: details are printed under the relevant figures.

At Sage, Robert Rojek has never faltered in his support, despite the seemingly endless apologies for not even starting the manuscript. I would also like to thank James Donald for his thought-provoking comments on a draft of the book and for his encouragement to go further.

None of this would have been possible without Jenny Robinson. Without her, nothing would be the same.

Steve Pile
London and Milton Keynes

Introduction
in which we explore how to explore
the phantasmagorias of city life

Introduction: what is real about cities?

Often enough, it can seem that what is real in cities is all the material stuff of life: buildings, infrastructures, money, labour processes, schools, housing, hospitals, consumption, and so on. This was not a mistake made by Robert Park, a leading member of the Chicago School. While attempting to define the essence of cities, he observed that:

> The city [...] is something more than a congeries of individual men and of social conveniences – streets, buildings, electric lights, tramways, and telephones, etc.; something more, also, than a mere constellation of institutions and administrative devices – courts, hospitals, schools, police and civil functionaries of various sorts. *The city is, rather, a state of mind,* a body of customs and traditions, and of the organized attitudes and sentiments that inhere in these customs and are transmitted with this tradition. The city is not, in other words, merely a physical mechanism and an artificial construction. It is involved in the vital processes of the people who compose it; it is a product of nature, and particularly of human nature. (Park, 1925a, page 1, emphasis added)

Robert Park was not impressed by the idea that the city could be defined solely by its physical or institutional forms. Nor is it simply a set of administrative devices that involve courts, hospitals, schools, the police, bureaucracies and city government. What is vital about cities is that they bring together people in such a way that this makes a difference to what goes on between them.[1] Whatever it is that makes a city a city, it has much to do with their social processes, their customs and traditions. The city does not just express itself in the buildings, the streets, the traffic that seem to define it, but in the ways in which people live, work, trade; their customs, habits, pleasures, crimes, angers. From this perspective, Park's statement that the city is a state of mind must be taken seriously.[2]

What makes the city a city is not only the skyscrapers or the shops or the communication networks, but also that people in such places are forced to *behave* in *urban* ways. For some, this involves an ever-increasing pace of life, the necessity of blocking out most of what goes on in cities, and a mental attitude based on calculation, rationality and abstract thought. In many ways, this implies that city dwellers are 'locked down' emotionally: reserved, detached, distant, calculating. On the contrary: I argue that *what is real about cities is the sheer expressiveness and passion of its life,* even in its most boring, or most objective, forms. What I would like to do in this book is play with the notion of what is considered real about city life. Not by specifying 'The Real' or by attempting to make 'The Real' more real, but instead by loosening and expanding both what we think is real about city life and also what we are prepared to take seriously in city life. For me, this means that more attention has to be paid to the city's state of mind.

Indeed, it can sometimes seem as if the city's state of mind — its sentiments, its attitudes, its sense of self, its mood — gives it a specific character all of its own. What is real about cities, then, is also their intangible qualities: their atmospheres, their personalities, perhaps. As graphic novelist Neil Gaiman observes,

> Each city has its own personality, after all. Los Angeles is not Vienna. London is not Moscow. Chicago is not Paris. Each city is a collection of lives and buildings and it has its own personality. (1993a, page 18)

Something about city life lends itself to being read as if it had a state of mind, a personality, as having a particular mood or sentiment, or as privileging certain attitudes and forms of sociation. It is quite clear that New York is not New Orleans, that London is not Singapore, that Paris is not Berlin. For sure, this has something to do with the buildings: with their built form, the super-structures and infrastructures of the city. For sure, it has something to do with the way people live their lives in cities, with their cultures and customs, with how they treat strangers, with their differences and indifferences. Yet, it is an odd thing to assume that cities have their own personality or state of mind. Surely, cities are far too chaotic and disorganised to be thought of like this; even those that are highly ordered. These commonplace experiences of the personality of a city may feel real, yet also they are phantasms that vanish as soon as light is cast upon them.

As I see it, this familiar experience of cities as being indefinably distinct from one another has something to do with the imaginary and emotional aspects of city life. In *Imagining the Modern City* (1999), James Donald makes a strong case for thinking about the 'structures of feeling' that comprise city life. He reveals these through explorations of a range of cultural phenomena, from cinema to city plans, to notions of urban citizenship and ethics. For him, the key to cities is that they have to be lived in, made habitable, *haunted* even. In Donald's work, as with my own, the analysis of the real of city life is expanded to include the shadows, irrationalities, feelings, utopianisms, and urban imaginaries. Donald makes clear that urban imaginaries are both emotional and unconscious, for example, when talking about the urban uncanny (1999, pages 69–73) or the city as a dream factory (pages 86–91).[3] For me, this implies that the images and representations of a city have much to do with how they feel, their personalities.

Perhaps these urban imaginaries can be collated, put together, to provide an account of the differences between — the individual personalities of — cities. However, this would be to overplay the 'individuality' of cities, as if they existed in splendid isolation from one another and as if the development of their 'personality' somehow owed nothing to their relationship to other places, other times. To get at the circulation of urban imaginaries, it will be necessary to track 'things' (by this I mean any object of consciousness, including fantasies, ideas, information, and so on) between and through cities. For sure, this will include images and representations, but as they connect to their emotional qualities.

However, revealing the emotional qualities of city life will also require an expanded sense of what is curious about cities, both their contents and their cultural expressions.

Boldly stated, I believe that greater emphasis in thinking about what is real about cities needs to be placed on the forms of *emotional work* that comprise urban experiences. These experiences might – if cities have their own distinctive state of mind or personality – be thought of as being structured in specific ways. A notion such as Raymond Williams' 'structures of feelings' (1973) promises much in revealing how a city – like an epoch – might have certain predispositions of sense and sentiment. The idea of structures of feeling implies an ordering or coherence to emotional life that can be characterised in specific ways, just as there might be a structure to personality. It also implies that it is expressed in various ways, in cultural forms such as literature. I would resist seeing this ordering as being preordained or fixed in stone or even orderly. Nor do I necessarily think that cultural forms of expression necessarily reveal, in some direct and obvious way, their feelings. Instead, these structures should be thought to be fabricated, devious, contradictory, mobile, changing and changeable. To evoke the more febrile, secretive and ambivalent aspects of emotional life, I prefer the term 'phantasmagoria'.

The term phantasmagoria implies many things. In some ways, it describes an experience of movement, of a procession of things before the eyes. In other ways, it invokes the importance not only of what can be seen, of the experience of the immediate, but also of life beyond the immediately visible or tangible. It suggests a quality of life that is ghost-like or dream-like. By evoking the dreaminess and ghostliness of cities, phantasmagoria is highly suggestive of the importance of particular kinds of emotional work for city life – yet these remain in question and require further exploration. Indeed, the emotional work that goes into these phantasmagoric experiences is far from well understood. Finally, phantasmagoria implies a peculiar mix of spaces and times: the ghost-like or dream-like procession of things in cities not only comes from all over the place (even from places that do not or never will exist), but it also evokes very different times (be they past, present or future; be they remembered or imagined). [I will return to the idea of phantasmagoria at the end of this introduction.]

What is real, then, about cities is as much emotional as physical, as much visible as invisible, as much slow moving as ever speeding up, as much coincidence as connection. In this light, it is best to think of this book as an exploration of urban phantasmagoric experiences, rather than as somehow laying bare the hidden 'Real' of city life. But how to do this? My approach to these questions relies on a combination of early Freudian psychoanalysis and Walter Benjamin's critical theory. This approach discloses the significance of dreams, magic, vampires and ghosts for the emotional work of city life. Dreams, magic, vampires and ghosts are significant in other ways too. They reveal aspects of the lived experiences of cities, of urban social processes, and also of the spatial and temporal constitution of city life.

I do not claim that the phantasmagorias of dreams, magic, vampires and ghosts account for the emotional labour of all cities, nor even for all the emotional work within any particular city, even those I have focused upon: that is, London, New Orleans, New York and Singapore. Instead, I have sought to open up a field of analysis that is capable of taking seriously the imaginative, fantastic, emotional – the phantasmagoric – aspects of city life. In doing so, I have found certain coincidences and connections within and between these cities. As a result, I have been able to foreground spatialities that are normally occluded in understandings of city life.

This approach, and the evidence that it has called into view, has not emerged in an intellectual or empirical vacuum. The broad intellectual heritage of this study might be described as psychogeographical. The attraction of this work has been its willingness to blend an appreciation of the social, physical, historical, spatial and psychological dimensions of city life into one. Further, psychogeographers have often sought to imaginatively reconstruct city life, piece by piece, through a style of observation that pays particular attention to the minutiae of the city. This imaginative reconstruction often entails the search for lost memories; memories that both interfere with the solidity of the present and also enable alternative futures to be imagined. My version of this is somewhat different from the psychogeographies that have gone before. Even so, it is still worth giving a flavour of these experiments – if only to convey the need to experiment.

The psychogeographies of novelist Iain Sinclair and film-maker Patrick Keiller, for example, explore the moods and atmospheres of London. Their work is not academic, and in many ways I think this is a better place to start an analysis of the emotional life of cities than with more academic accounts of, say, fantasy or affect. Psychogeography is a term most closely associated with the writings of Guy Debord and the Situationist International. Their investigations into, and interventions in, city life are worth taking into account because of the utopianism in their writings. Their contribution is to install the discovery of hidden or blatant power relations at the heart of psychogeographical practices. The intention was not simply to be able to describe the world in different ways, but to release people's Real desires and aspirations upon the world.

In much psychogeography, there is an underlying assumption that people's desires are prevented from gaining real expression and, instead, have become fetishized (or commodified) or blocked in various ways. This assumption that emotions are held at a distance, in some way, by city dwellers requires some explanation. Absolutely key to this is Simmel's writing on the city and mental life, so this is the next step to take. Simmel wanted to add a psychological dimension to historical materialist accounts of the city. His work provides a rich resource for thinking about the consequences of modern city life for affect, people's feelings and social interactions. Finally, however, we get to a place where we need to consider the relationship between the social and the personal, between the visible and the emotional. At this point, cities begin to shimmer: what is real is no longer quite where it was, even as a state of mind. In the final part of this introduction, we will think in more depth about the phantasmagorias of city life.

Before all this, however, let us look at Iain Sinclair's and Patrick Keiller's explorations of the emotional life of London.

Psychogeographies of London: Sinclair and Keiller travel in space and time

One way in which urban theorists have attempted to explore the overlooked or hidden aspects of cities is to travel across them: sometimes by train, sometimes by car, but mostly on foot.[4] For many, walking is the pre-eminent spatial practice for experiencing the city. In Iain Sinclair's novels, walking is significant not just as a means of experiencing the city and of assessing the mood of the city, but also of making deliberately imaginative – even mythic or magical – connections between different parts of the city.[5] Sinclair describes the advantages of walking like this:

> Walking is the best way to explore and exploit the city; the changes, shifts, breaks in the cloud helmet, movement of light on water. Drifting purposefully is the recommended mode, tramping asphalted earth in alert reverie, allowing the fiction of an underlying pattern to reveal itself [...] noticing *everything*. Alignments of telephone kiosks, maps made from the moss on the slopes of Victorian sepulchres, collections of prostitutes' cards, torn and defaced promotional bills for cancelled events at York Hall, visits to the homes of dead writers, bronze casts on war memorials, plaster dogs, beer mats, concentrations of used condoms, the crystalline patterns of glass shards surrounding an imploded BMW quarter-light window [...] Walking, moving across a retreating townscape, stitches it all together: the illicit cocktail of bodily exhaustion and a raging carbon monoxide high. (1997, page 4)

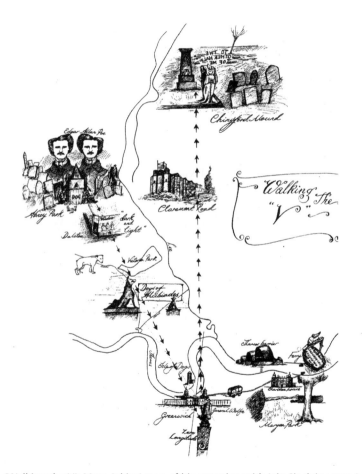

Figure I.1 'Walking the V': Marc Atkins' map of his excursion with Iain Sinclair across London.

In *Lights Out for the Territory*, Sinclair documents a series of nine excursions through London.[6] Sinclair's book, however, is not a simple account of these expeditions. Instead, he is attempting to write (what Sinclair calls) a psychogeography of the city that traces out the associations between the things that he notices and their lost or hidden histories.[7] By making purposeful associations between things, both geographically and historically, Sinclair hopes to provide an accurate assessment of the mood of London. Sinclair's project is of interest: partly because of the histories that he deploys in his narratives; and partly because of the nature of hidden connections criss-crossing the city that he attempts to divine. Sinclair describes the underlying logic for the first excursion like this:

> The notion was to cut a crude **V** into the sprawl of the city, to vandalise dormant energies by an act of signmaking. To walk out from Hackney to Greenwich Hill, and back long the River Lea to Chingford Mount, recording and retrieving the messages on walls, lampposts, doorjambs; the spites and spasms of an increasingly deranged populace. (1997, page 1)

By making this journey, Sinclair was trying to scratch a **V** on the face of London (see Figure I.1). More than this, however, by documenting every detail of the excursion, Sinclair was using the (not so) arbitrary **V** sign to make a cross-section of London's moods. Sinclair's own mood is spleenful. This becomes clearer as Sinclair follows his **V**-shaped path across London.

The excursion itself begins in Abney Park Cemetery, North London (Figure I.2). As he walks south from Stoke Newington towards Dalston, Sinclair takes the time to observe a stunted obelisk, which stands in front of the Duke of Wellington pub; an elderly man feeding pigeons with stale bread from bulging plastic shopping bags; a homeless woman slowly threading her way between lines of traffic during rush hour. He walks down Kingsland High Street, arriving eventually at Victoria Park which, for Sinclair, is too clean and tidy. Then, he walks south past the former 'The People's Palace' (which is now a college of the University of London) to the Isle of Dogs.

There, Sinclair disdainfully scrutinises Canary Wharf (Figure I.3), describing it as an extraordinary monument to Margaret Thatcher's virulent monetarism, and he nostalgically honours the persistence of Class War's graffiti, despite the decline in activities by the anarchist group. From there, Sinclair crosses the Thames into Greenwich. At Greenwich, he turns north to complete the second leg of the **V**. Finally, he arrives at Chingford Mount. Here, Sinclair concludes his narrative on two ironic notes: first, that suburban Chingford Mount cemetery is the final resting place of notorious East London gangster Ronnie Kray; and, second, that the cemetery is a dog free zone – Ronnie Kray loved dogs.

En route, Sinclair notes the almost ubiquitous graffiti that 'speaks' of the real feelings of the 'increasingly deranged populace'; he watches and participates in sinister events, such as the funeral cortège of Ronnie Kray through Bethnal Green; he remembers murderous deeds, such as the killing of 'Big Jim' Moody in an East End pub just off Victoria Park; and he searches for the ghost of a revolutionary politics in what might now be called the ruins of London. Sinclair is preoccupied with the seamier side of London's histories: its murders, its revolutions, its injustices. His London is mad, bad and sinister. I am less concerned with

Figure I.2
Broken angels in Abney Park Cemetery.

Figure I.3
A monument to Thatcher? Canary Wharf in the 1990s.

introduction

whether this is an accurate or useful description of London's mood or personality than with the effect of putting these apparently disconnected observations or events into the same narrative. As Sinclair threads his way through the occluded aspects of London life, the apparently unconnected stories assemble into a phantasmatic, surreal (over-real) urban scene. The stories show London life as a kind of phantasmagoria – a serial procession of hauntings ordinarily hidden from sight.

For Sinclair, the city is haunted by its misdeeds, so by tracing the stories of the dead he excavates the hidden violence underlying London life. In this ghostly city, we can see that many histories occupy the same space – and that these dark histories haunt Sinclair. For others, however, experiences of the city will be different. The point, here, is that there is a dense weave to the emotional fabric of the city, to the warp and weft of the city's histories and geographies. It is not Sinclair's intention to unpick this dense weave, but rather to uncover the patterns that lie underneath the visible city. Patrick Keiller's own psychogeography of London has similar intentions and methods. Like Sinclair's, Keiller's London is also phantasmatic, except his London is haunted by a London that never happened. Keiller's work is as much about the invention of memory as it is about really looking at the city.

In Patrick Keiller's film *London* (1993), an anonymous narrator tells the story of Robinson's experiences of, and thoughts and feelings about, the city. Like Sinclair's psychogeography, it relies heavily on journeys to piece together the city and also upon a lingering look that carefully notes what it sees. Like Sinclair, Keiller is interested in the almost forgotten histories that these journeys can bring to light. Keiller's attention to the detail of London life, though, is concerned with the search for locations of memory: memories that disturb the idea that the past necessarily had to lead to the catastrophe of the present. Keiller's film resonates with Sinclair's urban explorations, but there is also a sense of memory, of alternative histories, and of the potential for resignifying the city.

Figure I.4 London's neo-Georgian phone boxes.

Keiller's film is an attempt to examine the detail of the city to find its hidden secrets *in order to* unsettle assured histories of the present.⁸ For example, at a certain point, Keiller's film takes time to notice the fast-disappearing neo-Georgian telephone box (Figure I.4). At the time, London telephone companies were replacing these because, they said, they were hard to clean and expensive to repair when vandalised. Robinson, the narrator tells us, is characteristically nostalgic about the whole thing; something real was being permanently lost from public life, as phone booths were becoming more open to the streets and less able to be used as public conveniences. The smell of urine was being replaced by the smell of disinfectant.

Figure I.5
London's neo-Georgian phone boxes appear on cards posted inside London's neo-Georgian phone boxes.

Nevertheless, the neo-Georgian boxes have an after-life (Figure I.5). Some have survived, probably because they are so much a part of the distinctive look and feel (and smell) of London's streets. Indeed, they appear on many London postcards. Thus, fragments – such as these telephone boxes – become popular symbols for the city as a whole. In these fragments, arguably, can be found something of the attributes and affects of the city itself. The first step in understanding the city as a whole, then, would be to collect fragments of the city, like a crime scene investigator taking pictures of a crime scene from every angle. In Keiller's film, the narrator tells us that Robinson is making his own series of 'postcards' of London. These postcards, for Robinson, reveal the home-grown exoticism of London, as if he were collecting postcards from far-off lands. Indeed, Keiller's film *looks* like a series of postcards: each shot lingers on the urban scene, allowing it to be contemplated without distraction.

These postcards frame the city, slow it down, enticing us to look at the city with different eyes. By giving us the time to look at the spaces of the city, other aspects of the physicality of the city become apparent: its motion, its duration, its endurance, its persistence ... and its passing. The postcards become a procession of visual compositions, like a phantasmagoria. The narrator does not seek to interpret or to cohere these postcards. Instead, it is doubly detached: offering a narrative about a narrative,

Robinson's researches into London. The tone bears similarities to Sinclair's, but Keiller's film has its own atmosphere — a melancholic sense of a failed *present* pervades his film. The narrator is informative:

> The failure of the English revolution, said Robinson, is all around us: in the Westminster constitution, in Ireland. And poisoning English attitudes [...] Everywhere we went, there was an atmosphere of conspiracy and intrigue.

As this is said, the film shows MI6's (the British Secret Intelligence Service) dramatic building in Vauxhall.[9] London is an alchemy of intangible power, but also of the possibilities for an alternative present. Failed experiments in the history of the city can be discovered in its neglected spaces. Thus, the narrator describes Robinson's objective:

> He was searching for the location of a memory. A vivid recollection of a street of small factories backing onto a canal [voiced over a scene of waste land in Wapping, East London] but they no longer exist... And he has adopted the neighbourhood as a site for exercises in psychic landscaping, drifting and free association.

Expressions such as psychic landscaping, drifting and free association bear traces of both psychogeographical and psychoanalytic thought. The journey into memory is a dream-like one, attempting to capture those images and meanings that no longer exist or that might exist. Indeed, these exercises become ways of understanding the dream-like quality of cities: its scenes and associations, its juxtapositions and simultaneities, the compromises between the past and the present. Like a dream, the expedition consists of fragments of time and space, pieced together, given a sequence and a narrative by the journey itself. In these expeditions, the narrator explains,

> He seemed to be attempting to travel through time.
> [This is voiced over a prolonged shot of wooden cargo palettes burning, possibly/probably as a 'memory' of the Guy Fawkes night, seen later in the film.]

As Robinson conducts his search for a memory, it becomes clear that travelling through time involves travelling through the spaces of the city. Travelling, more precisely, is about moving in time-space: to find a memory, one also needs to locate it; to find its location is also to search out its past. By travelling in time-space, it is possible to reconstruct the city as it might have been. This means more than simply laying bare the city as it is or once was. It means placing the fragments of the city into a history that could have been, re-establishing their ties to other times, other possible presents, other futures. For example, Robinson angrily remembers that the urban waste lands in Wapping were created by the economic recessions of the late 1970s and 1980s. But these derelict spaces are not simply about emptiness, nor about the absence of history, nor the absence of anything going on. Such places, for Robinson, are full of ghosts who stir up a world where the jobs were saved and the people stayed.

For both Sinclair and Keiller, the journey is a deliberate attempt to experience the geography of the city differently. Keiller's film is a chronicle of Robinson's alternative times-and-spaces of London, each prompted by an expedition:

The first is to be a pilgrimage to the sources of English Romanticism. On March the 10th, we set out for Strawberry Hill, the house of Horace Walpole, but were disturbed by events on Wandsworth Common; the bomb had gone off at 7.10 that morning.

Many of the expeditions Robinson and the narrator set out on are overtaken by unexpected events; events created – we might say – by the city. Chance and coincidence are significant, for they allow for unplanned discoveries to be made. To begin with, Robinson's first journey is consistent with Romanticism itself: to see oneself from the outside, to see oneself as part of a romance. The film itself also involves the romance of origins and traditions. But these intentions are undone by the event. The Wandsworth Common bomb explodes the Romance, both of the expedition and the city. And the narrative is diverted into a consideration of how London is haunted by a not-so-post colonial war and its politics. The bomb provokes Robinson to reflect on Londoners' capacity to forget history – such as events in Ireland and the long-standing bombing campaign in London itself (Figure I.6). Thus, instead of discovering London's Romanticism, the urban explorers find themselves in a city that forgets both Romance and Terror – that even forgets History itself.

Figure I.6   Did an IRA bomb once tear these buildings to shreds?

If Sinclair's and Keiller's psychogeographies are a series of pilgrimages, they are less about reaching a source or a destination (the arrival at places already known) than about the amnesias, frustrations and diversions of the city. As the journeys are embarked on and undertaken, they trace out specific geographies and histories of the city. Sinclair and

introduction

Keiller emphasise particular events and focus upon particular surfaces of the city. Even so, in their work, it is possible to discern a way of thinking *from* fragments *to* a broader understanding of London and of city life. More interestingly, perhaps, each fragment suggests an alternative present – an alternative pattern to the city's histories and geographies. The times and the spaces of the city are no longer self-certain, each element is 'over-determined' (i.e. determined many times over): by memory, by meaning, by social relations.

The urban explorations of Sinclair and Keiller have three significant features: first, they allow the bits and pieces of the city to be stitched together in a variety of ways; second, they show how memories, some almost forgotten, can be made to flash up, disturbing the solidity of space and time; and, third, they illustrate the significance of the procession of visual, bodily experiences through city life. If Sinclair and Keiller follow particular paths in their books and films, and offer particular perspectives on what they see, this does not mean that they are the only ones. Instead, it might be worth following other paths, tracking other things, through the city. Importantly, their work shows the value both of using 'fragments' as starting points for these explorations and also of tracking 'things' through time and space. While the psychogeographies of Sinclair and Keiller convey a sense of the dream-like and haunted quality of city life, psychogeography – as conceived by Guy Debord and the Situationist International – also has radical intentions.

Psychogeography and the Real City

Guy Debord believed that practical experiments need to be conducted into city life. For Debord, these experiments would constitute new urban practices, which can be broadly labelled psychogeography.[10] Debord essentially describes psychogeography as a mode of observation. This mode was not meant to provide neutral descriptions of urban life, its intention was to scandalise: more specifically, to provoke a crisis in happiness. In part, this was to be achieved by delineating the human experiences of urban landscapes, showing where they seemed – but failed – to meet desire. At the heart of this idea is a sense that city life, modernity, had led desire up a cul-de-sac, where it promised satisfaction, but in fact met a dead-end. Through a variety of direct interventions in urban space – the creation of situations – ways were sought out of these cul-de-sacs.

A central (revolutionary) task for psychogeography was to map people's experiences of the city. In order to produce this map, new ways of moving through the city, and of charting its diverse ambiences, would be needed. One such technique was the *dérive*, or *drifting*. Drifting was not a random activity, but a deliberate attempt to think spaces and spatiality differently: to create situations in which space as currently produced comes to be seen as absurd. Thus, the *dérive* is

> a technique of transient passage through varied ambiences. The dérive entails playful-constructive behaviour and awareness of psychogeographical effects [...] from the dérive point of view cities have a psychogeographical relief, with constant currents, fixed points and vortexes which strongly discourage entry into or exit from certain zones. (Debord, 1956, page 22)

For Debord, these spatial practices showed that the city was divided into distinct atmospheric zones, with distinct ambiences (and micro-climates). Paradoxically, moving through the city in these playful, disruptive ways demonstrates the settled geographies of power relations in the city:[11] the places where access is denied – from the gates of government institutions, to military sites, to buildings for spies, to private houses. The city is more closed than open: porous, because of what it filters out and what it channels through specific routes, rather than because it allows things to flow free or lets everything through. Moreover, the freedoms of the city appear to be constantly under attack in the modern city, constantly circumscribed, constantly surveilled – often enough in the name of freedom, service and protection.

Psychogeography as practised by the Situationist International (SI), and also by Keiller and Sinclair, is the study of immediate experience of the environment – the city – and its impacts upon human emotions. The *dérive* is a technique through which these emotions are to be registered and experienced. Thus, psychogeographers survey the city, taking notes on the atmosphere, ambience, mood, mode of feeling, at any given place. But they are also attuned to the *power relations* that underlie these experiences. For the Situationist International, these power relations – often implicit in the production of space, of the city, of commodities, of modern life – needed to be *turned* towards other ends. This turning, or *détournment*, does not mean there is a destination or goal state. Instead, *détournment* instigates a search for points of departure.

Psychogeography was turned towards desire, towards excitement, towards life.[12] Its implacable enemy, boredom: 'Boredom is Counter-Revolutionary', the SI slogans shouted.

> We are bored in the city, there is no longer any Temple of the Sun [...] We are bored in the city, we really have to strain to still discover mysteries on the sidewalk billboards, the latest state of humour and poetry [...]. (Gilles Ivain [Ivan Chtcheglov], 1953, page 14)

Figure I.7 The billboard tells us how the new Lexus will change everything. The graffiti says, 'especially the climate!'

Ivain then lists some billboard captions (like that in Figure I.7). And some places: buildings on streets, something he has noticed, something strange, perhaps a memory, something 'without music and without geography' (page 14), and something that you cannot see: 'it does not exist' (page 14). He continues:

> All cities are geological; you cannot take three steps without encountering ghosts bearing all the prestige of their legends. We move within a *closed* landscape whose landmarks constantly draw us toward the past. Certain *shifting* angles, certain *receding* perspectives, allow us to glimpse original conceptions of space, but this vision remains fragmentary. (Ivain, 1953, pages 14–15)

A new formulation of city life would have to attend to the fragments, the pasts, the ghosts, the magic of the city. Keiller's *London* bears some of the mode of feeling of Ivain's analysis of the city, especially in its sense of a closed landscape, its mourning for something lost, its shifting angles of perception, its collection of fragments. Most of all, they share a sense that the city can be explored through even its tiniest details or in its surface appearance: as Keiller's narrator intones

> Robinson believed that, if he looked at it hard enough, he could cause the surfaces of the city to reveal to him the molecular basis of historical events and, in this way, he hoped to see into the future.

In these psychogeographical works (by Keiller, Sinclair and Ivain), it is possible to see an argument about the *real* nature of the city. As a series of postcards of the city, Keiller's *London* asks the audience to (really) look at the city: to see beyond its superficial appearances, its surfaces.[13] Through the series of postcards, Keiller hopes to expose the links between tiny, molecular events and the grand flow of history. There is a strong sense of the multiple and overlapping histories and geographies that make up (both comprise and invent) the observed, visible cityscape. As we sit and watch the cityscape from a certain angle, traffic, people, events pass before us – time itself passes – all following their own paths.

Figure I.8
The gates of Victoria Park, East London.

In Keiller's film, the gateposts of a park that Robinson visits are bearers of memories, of the sounds of children (like those in Figure I.8). They are due to be knocked down – to become the ghosts that Ivain talks of. Listening carefully, Robinson can hear the gateposts with the sounds of what is about to be lost. The gateposts already haunt the city, ghosts from a future yet to come. By looking hard at the city's surfaces (for

Robinson), by noticing everything (for Sinclair), by encountering ghosts (for Ivain), it might be possible to discern patterns of historical events, as well as the many futures of the city, *beyond the surface appearance of things*.

These psychogeographies are evocative of something, but what? Of ghosts and dreams? Of pasts and futures? Of promise or failure? But surely these are not the only things going on in the city. Other stories are possible too. Behind the postcards, we can see the writings that tell other kinds of story – family romances, perhaps: 'having a lovely time'; 'wish you were here'. Behind the surfaces, perhaps we can also make out other (hi)stories of the city. It is the porosity of these urban spaces that evokes *both* the multiplicity of stories *and also* the many time-spaces of the city, only some of which are allowed to become real. Others become ghosts; some remain dreams.

The intention of psychogeographical experimentation and intervention was to explore the different ambiences of the city. It sought not only to map them, but also to turn them into something less settled, less alienated. However, these writings also revealed the city to be phantasmagoric: a spectacle, where the surfaces of the city did not betray the means through which they were produced. As important, the city was full of movement, of crisscrossing times and spaces, a serial procession of dreams and ghosts. The unresolved puzzle of the city lies in its shifting perspectives, in the way it shifts perception. In this urbanised space, we can no longer expect to find one answer, or one dream, as if it pointed to only one meaning. Instead, these psychogeographic experiments convey a sense of the multiplicity of cities, that overlap, pass by one another – that cross, get crossed and get cross.

Though there is a clear intention to explore the atmospheres and feelings of the city, no view is presented of the relationship between the city and psychological life. Nonetheless, there is a certain knot in this structure of feeling, tied between boredom and excitement, antagonism and indifference. While there is an appreciation of the map of feelings in psychogeography, we need to turn to Simmel, because he offers a way to understand the interrelationships between the atmospherics of city life and the states of mind of city dwellers.

The City and States of Mind

Simmel's classic essay 'Metropolis and Mental Life' (1903) explores how people's experiences of metropolitan life have changed, in comparison both to the past and to rural life. In this account, the street is seen as the place where people had to cope with the sheer vibrancy and diversity of urban experiences. Urban dwellers, with this in mind, develop a resistance to the sheer volume of those experiences: they become more aloof, more intellectual in the way they deal with the world around them (see also Allen, 1999a). Before we look more closely at Simmel's interpretation of the mental life of cities, it will help if I give a little background to Simmel's approach to social analysis. In the preface to his key work, *The Philosophy of Money* (1900), Simmel argues that, in the book,

> The attempt is made to construct *a new storey beneath historical materialism* such that the explanatory value of the incorporation of economic life into the causes of intellectual culture is preserved, while these economic forms themselves are recognized as the result of more profound valuations and currents of psychological or even metaphysical pre-conditions. (page 56, emphasis added)

This statement of purpose also describes the approach Simmel takes in 'Metropolis and Mental Life' to the problem of city life. Simmel is attempting to understand the psychological preconditions of modernity *while at the same time* explaining the development of those preconditions with reference to changes in the metropolis. In other words, Simmel is arguing both that psychology underpins economic life and city life, and also that changes in both economic life and city life alter people's psychology. It is best to think of Simmel as examining how different aspects of city life – the economic (mainly in the form of money) and the psychological (in the form of attitudes and associations) – *relate* to one another, where each aspect of city life is capable of modifying any other.

Simmel's view that something significant was happening to the mental life of urban dwellers is clear from the outset of 'Metropolis and Mental Life':

> **The deepest problems of modern life flow from the attempt of the individual to maintain the independence and individuality of his [sic] existence against the sovereign powers of society, against the weight of historical heritage and the external culture and technique of life. (1903, page 30)**

Simmel is suggesting that the deepest problems for urban dwellers stem from their attempts to cope with the pressures of modern society, culture and techniques of living. It is also clear that Simmel does not believe that such pressures existed prior to the eighteenth century, nor moreover do they exist in rural areas. In other words, Simmel's analysis of the changing relationship between the metropolis and mental life is situated within an historical account of the changing nature of cities. Basically, he is suggesting that cities have become the sites of liberation from traditional ties, whether economic, moral or religious. Simmel argues that modern urban dwellers struggle to maintain their sense of self, while being severed from (or liberated from) traditional ties and while becoming embroiled in mass society and the money economy. Thus, the modern city sets quite

Figure I.9 Piccadilly Circus in central London.

different problems for urbanites from those that had existed either in earlier times or in the countryside. These problems have to do with the rise of the money economy, for sure, but they are also bound up with the sheer quantity of things going on in cities.

The feel of the big city was different and this was visible in its street life (Figure I.9), where there was a seemingly endless stream of things passing-by, as in a phantasmagoria. Cities were different, moreover, because they were constantly throwing people into contact with new experiences, new situations, and new people. Metropolitans responded to the possibility of being overwhelmed by all the new things they encountered by developing a particular set of attitudes – an urban mentality. This psychological structure is characterised by the attempt to dampen down the emotional intensity of urban stimuli. The metropolitan personality is founded on an *intellectualism*, as individuals seek to reduce their emotional responses to the volume and variety of stimuli that cities offer. Thus, Simmel is arguing that the growth of cities has a direct impact *both* on social interactions between people *and also* on their personal attitudes and behaviours:

> **Instead of reacting emotionally, the metropolitan type reacts primarily in a rational manner, thus creating a mental predominance through the intensification of consciousness [...] It is in this very manner that the inhabitant of the metropolis reckons with his [sic] merchant, his customer, and with his servant, and frequently with the persons with whom he is thrown into obligatory association. (Simmel, 1903, page 32)**

For Simmel, the characteristic rationalism and intellectualism of the urban dweller has far-reaching implications. In some ways, these predispositions allow urban dwellers to shake off older customs and traditions of social interaction. The modern city, Simmel argues, creates opportunities for people to develop a new sense of their individuality and uniqueness. Now, this is not simply a psychological process, it is also related to changes in the economy: for the economy now values personal creativity and capacities. But Simmel is not content only to look at the ways individuals are liberated in cities, for he also identifies how these new freedoms create new kinds of problems for people. In the city, people are bound up in all kinds of social interaction, but these are more and more with people with whom they have only the most superficial relationships.

The essence of the modern city is the way it concentrates – *in time and space* – many activities, people and things (including political institutions and money). As a result, people can withdraw emotionally from city life and develop a sense of reserve. However, the concentration of stimuli can also cause people to become blasé about, or indifferent to, things. By this, Simmel meant that people no longer cared so much about the differences between things. This attitude, for Simmel, was associated with the rise of commercialism and the money economy. Effectively, money made everything purchasable and exchangeable. The value of things could be calculated exactly: more and more, Simmel argued, city inhabitants developed a 'calculating exactness of practical life' (1903, page 33).

Reading Simmel as an eye-witness (rather than a theorist), we can see that, in the early part of the twentieth century, European cities were changing in ways that were having *noticeable impacts* both on the organisation of society and on the ways people responded to each other. That is, Berliners at this time were developing new ways of relating socially to one another. These new ways of relating, significantly, involved the reordering of time and space. Punctuality – being in the right place at the right time – had now become important. Pocket watches were visible evidence of this changing significance of time. In fact, pocket watches had become highly fashionable in Berlin during that period.[14] Urban life was now being divided into minutes and seconds, rather than hours. Space, too, was being made in different ways. Flows of things through the city were being controlled in ever more exact and integrated ways. Emblematic of this control over space was the traffic signal, according to Simmel. Meanwhile, the land was being divided into different properties, with clear distinctions between public and private space.[15] Money, also, was changing the way that property (i.e. ownership of space) was viewed within cities, such that higher profits were now being generated within the commercial areas of the city.

> The metropolis has always been the seat of the money economy because the many-sidedness and concentration of commercial activity have given the medium of exchange an importance which it could not have acquired in the commercial aspects of rural life. (1903, page 32)

Though the city had always held the purse strings, Simmel thought that increasing agglomeration in modern cities was making a qualitative difference to people's ways of life. The temporal and spatial concentration of modern cities had intensified the relationship between people to the degree that they now had to hold the world at a psychological – and physical – distance. One might think, here, of the way the telephone allows people to talk to one another at great distance, but it also absolves people of the need to meet face-to-face or to involve themselves personally with those they talk to. Simmel's city is a double-edged sword: its up-side is often also its down-side.

Simmel remained cautious about the possibility that the freedoms of the city were necessarily a good thing, nor was he optimistic that, by gathering people of different backgrounds into close proximity and into social interactions, the city would promote harmony and mutual understanding. Indeed, the seeming coldness and unfriendliness of city dwellers had a much bleaker side. Simmel worried that urban dwellers also exhibited for each other 'a slight aversion, a mutual strangeness and repulsion which, in a close contact which has arisen anyway whatever, can break out into hatred and conflict' (1903, page 37).

For Simmel, the emotional life of cities is distinct from life to be found elsewhere. Central to his analysis is a sense of the diversity and heterogeneity of urban experiences. The endless procession of these experiences creates the city as a phantasmagoria. What Simmel adds to our understanding of the city is a psychological dimension, in which urban dwellers develop a series of attitudes that enable them to cope with the pressures of modern city life. These include the blasé attitude, but also reserve, indifference, calculativity, rationality, individuality, personal skills, creativity, specialisation and the like. Urban etiquettes and manners are different, but so too are the production and the social use of technology. Nevertheless, there is no reason to assume that all cities carry the same etiquettes and uses of technology, nor that all people in cities necessarily respond in the same way.[16]

Modern city life is characterised, in many ways, by its phantasmagoric experiences. From Simmel, we learn that these experiences induce, and are produced by, psychological and emotional reactions in modern city dwellers. The phantasmagorias of modern city life are, to a degree, created by the sheer variety of things going on there: in part, this has to do with social processes, including for example the state of the money economy or commercial activity or technological development; in part, it involves the inability to really grasp all the different things that are going on in the city, or even the means through which city life is produced and reproduced; in part, it entails how people express and suppress affect.[17] These are part and parcel of the dream-like, haunting qualities – the phantasmagoric make-up – of city life. In this understanding, the idea of phantasmagoria is critical for an analysis of real cities, so it is to this idea that we turn next.

Phantasmagoria and the Realness of City Life

So far, what has been important is a form of attention to city life that is able to take seriously things that have been overlooked, discarded, ignored, forgotten, and so on: whether these are tall buildings or telephone boxes, gravestones or statues, graffiti or billboard captions. Paying this kind of attention to city life is a way of seeing the differences between cities and within cities, even while looking at apparently similar 'things' (such as monuments, buses, shopping malls, electric lighting and so on). The surface appearance of city life – whether it is in Sinclair and Keiller, the Situationist International, or Simmel – has turned out to be problematic in various ways. It is not that anyone is suggesting that the surface appearance of city life is not of interest, or even fundamental to understanding city life. It is more the case that these writers are suspicious both of what lies before their eyes and also of the consequences of being immersed in this world of sights. While Sinclair attempts to discern the secret or mystical connections between the things that he notices (whether these are lay lines, or a criminal or revolutionary underground), Keiller attempts to see the historical patterns in the molecular events that comprise city life. Debord, meanwhile, was suspicious of the ways in which society was becoming experienced through images, through the spectacle of life in the city. And Simmel wanted to know how it was that these shallow surface interactions structured the feelings of urban dwellers and the affect of city life.

Even so, the surface appearances of city life are pleasurable and distracting. And they will have other qualities too. To get at some of these, the term phantasmagoria becomes useful. It is a term most often associated with the work of Walter Benjamin, who occasionally suggested that the many surface appearances of the city gave it a dream-like and/or ghost-like quality. To understand why Benjamin was interested in phantasmagorias, the best place to start is with the technology itself. Crary tells us that a phantasmagoria was a specific form of magic lantern performance, popular in the 1790s and early 1800s (1990, page 132). It used back projection so that the lanterns were invisible to the audience: that is, such that the processes underlying the production of the projection were imperceptible. For Terry Castle (1988), the technique of phantasmagoria created the possibility for new visual experiences. Part of this had to do with the movement of images. Without a sense of how the images are produced, the transition between one image and another created a dream-like and ghost-like visual effect.

Moreover, in the nineteenth century, a whole new variety of new forms of image projection were being put on display. Benjamin lists just some:

There were panoramas, dioramas, cosmoramas, diaphanoramas, navaloramas, pleoramas (*pleo-*, 'I sail', 'I go by water'), fantoscope<s>, fantasma-parastases, phantasmagorical and fantasmaparastatic *expériences,* picturesque journeys in a room, georamas, optical picturesques, cinéoramas, phanoramas, stereoramas, cycloramas, *panorama dramatique*. (1927–40 [Q1,1], page 526)

All of which makes DVD and CGI look short on imagination. What is key is that phantasmagoria both consisted of a procession of images that blended into one another, as if in a dream, and also concealed the means of its production. In using the term, Benjamin was simultaneously evoking both aspects of the experience of phantasmagoria: both the spectacle of movement and the requirement to look beyond surface appearances to the means of their production. Thus, by using the term phantasmagoria to describe modern city life, Benjamin is highlighting both the new forms of experience visible in modern cities, such as Paris, and also the invisibility of the social processes that create the city's many spectacles.[18] For example, crowds had a phantasmagoric effect. Faceless people endlessly crossing before the eyes, one blending into another. The effect was alienating (as Baudelaire describes): partly because the technologies producing crowds remained out of sight, partly because of the impossibility of making contact with the ghost-like people in the crowd.

Alongside a sense of the magical production of visual experiences, and of the dream-like and ghost-like movement of figures, phantasmagoria does another kind of work for Benjamin. For, it is also an analytic concept that describes the nature of what he called the dreaming collective.[19] As such, it encompasses an appreciation of the general condition of modernity in which people sleepwalk their way through their lives, unable to wake up to their desires. But it also referred to the specific sites where this dreaming collective is produced and reproduced: for example, in sites of distraction or in the way that people protect themselves from the shocks of city life (as Simmel suggests).

Indeed, phantasmagorias were playing to – and creating – new forms of experience, including new forms of distraction and of boredom. In Paris, for example, the crowds, the streets, fashion, walking, journalism, detective stories, world fairs, offices, interior decorating, commodities, old photographs, and the like, created a variety of phantasmagoric experiences. Not only was the real life of cities, such as Paris, phantasmagoric, Benjamin suspected that modernity itself was phantasmagoric:

The world dominated by its phantasmagorias – this, to make use of Baudelaire's term, is 'modernity'. (Benjamin, 1939b, page 26)

Modernity is its phantasmagorias. That is, images of modernness and progress have turned the city, and even history itself, into a phantasmagoria. Critically, these phantasmagorias – the phantasmagoria of modernity itself – created the possibilities for new forms of social life. Like Simmel, Benjamin was not lured into thinking that these experiences were simply new. His suspicion of the ever-changing was that it concealed something that was ever-old. As such, the deceptions of phantasmagoria even extend to the flow of history itself:

The dreaming collective knows no history. Events pass before it as always identical and always new. (1927–40 [S2,1], page 526)

From this perspective, the modern city could be characterised by the experience of procession and movement, of appearance and spectacle, of progress and the new, of dreams and ghosts. But one in which the creation of new things, such as commodities, and the discarding or destruction of old things (such as obsolete commodities), doesn't necessarily imply progress, or even that people were getting (closer to) what they (really) *want*. Modernity is, in this light, inherently contradictory: new, yet old; offering satisfaction, yet constantly disappointing. Modernity is a magical process in which, as Marx and Engels famously observed, 'all that is solid melts into air' (1872, page 83).[20]

For Benjamin, these experiences of modernity created a kind of phantasmagoric dream-world. The idea of a dream-world is more than a description of modernity and the city, however, as it also has radical implications. From Freud, Benjamin takes the idea that real desires and wishes are wrapped up in dreams. For both of them, dreams are a way to understand these (hidden) desires. From Freud, Benjamin learnt that desires are disguised in dreams by processes of transformation he called 'dream-work', so understanding the dream-world would require understanding the work that had gone into producing them. As with dream-worlds, so it is with phantasmagorias. The injunction, then, is to explore the *work* that goes into making the phantasmagorias of modern city life. The logical place to start such an analysis is with the dream itself: with the work that goes into making dreams and with how this casts light upon the phantasmagorias of modern cities.

The first step I will take in exploring the phantasmagorias of city life is (as logic suggests) dreams, but dreams will lead to other urban phantasmagorias. Thus, the idea that dreams contain wishes, for example, will be carried forward into an exploration of magic in cities. Magic itself presumes the kinds of supernatural connection that Sinclair was seeking to discover in cities. The supernatural does not just suggest other forms of power, it also implies that there is more to life and death than life and death. In cities, this raises questions about the dead and the undead. At this point, the vampire becomes a key figure in thinking through modern city life. Vampires reveal something of the heterogeneous, non-linear, times and spaces that comprise cities. It is in this chapter, that the consequences of thinking about phantasmagorias of procession, or circulation, are most explicitly worked through. A discussion of the undead vampire leads us (full circle) to ghosts of the dead: that is, the procession of ghost-like figures in cities. Each chapter explores the ways in which the phantasmatic works, differently, to produce real city life. Throughout, dreams, magic, vampires and ghosts are seen as figures that simultaneously reveal and conceal desires and anxieties, often ambivalently, always contingently.

The exploration of the emotional life of cities begins, then, with *dreams*. Dreams do triple duty in this book. First, dreams can be found in cities, as fragments: in billboards, in stories about cities, in people's experiences of cities, and the like. Also, cities can be found in dreams. And this is not an inconsequential observation. Dreams, for me, are empirical phenomena found in both sleeping and waking life: found in many cities, found in certain sites, and so on. Second, dreams provide a model for interpreting city life,[21] for exploring the dream-like aspects of phantasmagoria. Freud's interpretation of dreams is revisited to provide an understanding of the work that goes into making dreams. This method of interpretation proceeds by following networks of feeling and meaning, beginning with specific dream elements. Dream analysis can then be used to understand both the dream-like qualities of city life and also the work that goes into producing the phantasmagorias of city life – not just dream-work, but also other forms of emotional work too, such as magic-work, blood-work and grief work. Third, dreams can be used rhetorically: they call to mind a vision of a better life, a better city. Dreams, in the end, are also political – as we will see in the conclusion to the book.

As a model of city life, dreams respond to Sinclair's injunction to 'notice everything'. In fact, it's presumed, in a dream analysis, that seemingly marginal aspects of city life may well be the most telling, the most revealing. In this book, I have focused on the urban marginalia of dreams, magic, vampires and ghosts – yet these will turn out to lie at the very heart of what is real about city life. What is interesting about allowing these marginal figures to be important is that they reveal the desires and fears of cities, and they do so differently in different places. So it is worth noting that other marginal figures can be of significance (and this book is an implicit demonstration that these should be taken seriously too). For example, certain kinds of rambler in cities were described as 'werewolves', because of the way they changed once they stepped out of their front (or back) doors.[22] They were Jekyll and Hyde characters,[23] who became more bestial, hairier, once loose in the social wilderness of the city. The werewolf also shows what happens to the city after dark, especially on a full moon![24]

Indeed, legends are tied to the very foundations of some cities. In Lund, Sweden, the story goes that a giant built the cathedral there (Figure I.10).

Figure I.10
Lund Cathedral was built by a giant.

Figure I.11 Ghost rats have been seen running from this raised mound in Lund.

Also in Lund, a werewolf is said to have roamed the small wood by the cathedral, while ghost rats have been seen coming from what looks like a burial mound (Figure I.11).

Angels, meanwhile, are said to have founded many cities. I am told that this is so common in Brazil it is unremarkable. In Budapest, there is a statue of the Archangel Gabriel in Heroes Square because (in one version of the legend) the angel is said to have appeared in a dream to Stephen and offered him the crown of Hungary (Figure I.12). Meanwhile, the city's catacombs

Figure I.12 The Archangel Gabriel looks down on Heroes Square, Budapest.

Figure I.13 Statue Park, Budapest, ghosts of previous regimes.

Figure I.14 Budapest is a city of angels.

and Statue Park suggest the ghostly traces of previous eras (Figure I.13).[25] In fact, wherever you look in Budapest, there seem to be (missing) monuments and golden angels (Figure I.14). In other cities, figures include the sandman, zombies, bogie-men (of various kinds), demons (especially, paradoxically, in Los Angeles, the city of angels), aliens and so on.

Surely such things are suggestive of how cities work, emotionally, imaginatively, fantastically, really. In this sense, crudely, one might intuitively surmise that a city covered in angels might have a different structure of feeling than one whose legends refer to Voodoo.[26] Following dreams, however, has led me to specific phantasmagorias and phantasmagoric figures. I have limited my study to dreams, magic, vampires and ghosts. I have followed a logic that takes me from one to the other, partly because they are also closely interconnected. This logic is explained in the conclusions and introductions to each chapter. Tracking these phantasmagorias and phantasmagoric figures has taken me to London, New Orleans and Singapore – in which dreams, magic, vampires and ghosts all appear in some form. Even so, as we will see, most are to be found in other cities also, such as New York, Johannesburg, Berlin and Paris.

In the next chapter, I explore in more detail the dream-like quality of city life. As a model for interpreting city life, it emphasises the changing meanings of 'things', and relies on tracking 'things' to bring new meanings to light. The idea that work goes into making dreams is highly suggestive, also, for thinking about the work that goes into making, as Benjamin might have it, the specific phantasmagorias of city life and the phantasmagoria of modernity. Let us now turn to the dream-world of city life.

1
The Dreaming City
in which cities turn into
dreams and dreams turn into cities

> The reform of consciousness consists *entirely* in making the world aware of its own consciousness, in arousing it from its dream of itself, in *explaining* its own actions to it. (Karl Marx, 1843, in a letter to Arnold Ruge)
>
> Every epoch, in fact, not only dreams the one to follow but, in dreaming, precipitates its awakening. (Benjamin, 1935, alternative translation, page 13)

1.1 Introduction

In his novel *Invisible Cities* (1972), Italo Calvino imagines a meeting between Marco Polo and Kublai Khan. In the course of their conversations, Marco Polo conjures up images of many fabulous and incredible cities. At one point, however, the Great Khan interrupts Marco Polo and begins to describe a wondrous city that he has thought of. The Khan then asks Marco Polo whether it exists. But it seems he had not been paying attention, for Marco Polo says he had been telling the Khan about precisely that city. In disbelief, yet also intrigued, the Khan demands to know the name of the city. Marco Polo replies:

> 'It has neither name nor place. I shall repeat the reason why I was describing it to you: from the number of imaginable cities we must exclude those whose elements are assembled without a connecting thread, an inner rule, a perspective, a discourse. With cities, it is as with dreams: everything imaginable can be dreamed, but even the most unexpected dream is a rebus that conceals a desire or, its reverse, a fear. Cities, like dreams, are made of desires and fears, even if the thread of their discourse is secret, their rules are absurd, their perspectives deceitful, and everything conceals something else.'
>
> 'I have neither desires nor fears,' the Khan declared, 'and my dreams are composed either by my mind or by chance.'

'Cities also believe they are the work of the mind or of chance, but neither the one nor the other suffices to hold up their walls. You take delight not in a city's seven or seventy wonders, but in the answer it gives to a question of yours'. (Calvino, 1972, pages 37–38)

In Calvino's hands, the city is much like a dream, a wondrous puzzle with a hidden secret. This understanding of the dream has much to offer an interpretation of city life and its atmospheres. Through Marco Polo, Calvino indicates that the meanings and feelings of city life are a rebus, in part because component images of the city shift around to produce new patterns that both illuminate and confound at one and the same time. The rules governing the enigma of city life, however, remain opaque. According to Marco Polo/Calvino, despite the opaqueness of these rules, there is a hidden secret to any city: that is, underlying the production of cities, there are the hidden workings of desire and fear. In other words, cities are desire and fear made concrete, but in deceitful and disguised ways. It is the same with dreams. Cities are like dreams, then, both because they conceal secret desires and fears and also because they are produced according to hidden rules that are only vaguely discernible in their surface appearances.

Kublai Khan cannot accept this interpretation, either of cities or of dreams. Surely, dreams and cities have nothing to do with one another. Dreams are illusions, unreal. Cities are very real, the work of the conscious mind, permanent in streets and spires – not the random, absurd juxtaposition of astonishing images.

Dreams would at first sight appear to be a phenomenon of the individual, of the sleeping mind, disconnected from the waking world. Dreams are personal, an arbitrary collection of irrational, absurd, puzzling, purposeless, evanescent impressions. On the other hand, cities are collective enterprises, founded in conscious, rational, intentional, purposive, durable, reflexive, practical activity. It is not just that cities are not like dreams, but that they are the products of entirely separate worlds: one sleeping, one waking. The important thing is not to jump too quickly to the conclusion, as Kublai Khan does, that dreams and cities belong to separate worlds. For dreams and cities are delightfully similar puzzles.

In this chapter, I will treat the worlds of dreams and cities, the waking and sleeping worlds, as interrelated puzzles. I argue that Freud's interpretation of dreams offers a model for understanding the thought and work that goes into making both dreams and cities. In this model, dreams will act as a bridging point between two other (often severed) worlds: the personal and the social. Dreams are the product of an individual's sleeping mind, but there is also a pervasive social discourse of dreams, whether this is found in political rhetoric or in advertising captions. Dreams are very much part of the waking world of cities. Indeed, the social discourse of dreams is so

Figure 1.1
Dream holidays for cold, tired underground Londoners.

Figure 1.2
'Long Live Dreams' says American Express, Oxford, England.

Figure 1.3
Liz Claiborne knows 'what dreams are made of', New York.

pervasive that it is itself phantamagoric. Dreams become part of the constant procession of 'things' through the city — dreams constantly pass before the eyes (see Figures 1.1, 1.2, and 1.3).

The forms that the phantasmagoria of dreams take requires some investigation, so I will describe various instances of the appearance of dreams in cities in **section 1.2**. This shows some of the key ways that the discourse of dreams helps to create a dream-like experience of city life. The intention here is to show that dreams matter in everyday waking life and that they act as a point of contact between social and personal worlds.[1] It is also the case that cities appear in dreams and, indeed, in stories about dreams. In **section 1.3**, I examine two stories taken from Neil Gaiman's classic *Sandman* series: each is a story about a city in a dream. Both tales are revealing of the uncanny feelings that can be provoked while thinking about dreams and cities, especially when they are in danger of vanishing or, indeed, awakening.

Moving between dreams and cities suggests that more attention be paid to the work that goes into making dreams. Indeed, Benjamin's injunction (as we saw in the Introduction) is to look beyond the surface appearance of images, such as dreams, in the phantasmagorias of city life. So, in **section 1.4**, I focus on Freud's account of dream-work, but with special interest in the key processes of condensation, displacement, representation and secondary revision. Together, these ideas suggest that thoughts and affect are shifted along trains of thought in various ways, constantly arriving and leaving, constantly being worked and reworked. As with dreams, so it is with cities.

Freud's specification of dream-work points to the work that goes into making urban space, and this is spelt out in **section 1.5**. There is an irony here. Dreams are part of the phantasmagorias of city life — that is, part of the imagery intended to disguise its own production — yet understanding how dreams are produced offers clues both to how urban phantasmagorias are produced and also to how these phantasmagorias are experienced, emotionally. First, elements or fragments in cities point not to one emotional content but to many. Second, cities are made up of heterogeneous elements that are 'overdetermined'; that is, they have an intensity of meaning and affect that is produced many times over by many different agencies. Third, the paths through which fragments of the urban scene come to be there are significant: partly because of the transformations that fragments go through on their way to the city and, again, once they are there; partly because of the transformations they create, as they go and once they get there; and, partly because of the (often disguised or displaced) intensity of feelings bound up in these fragments and their transformations.

The city is constantly churning things, people and ideas. Yet, this does not necessarily produce desired-for outcomes, nor outcomes closely linked to people's wishes, nor even fulfilling people's wish that wishes be fulfilled. This idea is explored using Walter Benjamin's use of dreams in his analysis of modernity and the city, both as a way to evoke a quality of life, and also as a revolutionary technique. To think this through, in **section 1.6**, I look at Walter Benjamin's analysis of the simultaneous dreaminess and derangement of city space. While Freud's understanding of dreams informed Benjamin, a Freudian re-reading of Benjamin's work provides a greater sense of the interweaving of social and personal worlds, and of the worlds of dreams and cities. Significantly, and as Benjamin would wish, dreams represent a breaching point between the personal and the social which opens up possibilities for re-imagining city worlds. So, in **section 1.7**, I examine Walter Benjamin's suggestion that people sleepwalk through modernity and modern cities, as if in a dream. According to Benjamin, moderns are in thrall to the phantasmagorias of city life. Yet, for Benjamin, there is still hope, as dreams anticipate awakening from the nightmare of modern city life.

Now, let us witness a procession of dreams, of dreams as commonplace images in the everyday life of cities.

1.2 A Phantasmagoria of Dreams?

The question for this section is whether dreams pervade everyday life, forming a chaotic procession of images in space and time. To answer this question, I present a wide range of examples of the ways in which dreams appear in waking life. The sheer number and variety of these examples goes a long way to establishing the case that there is indeed a phantasmagoria of dreams, where each instance of dream in the phantasmagoria provides an opportunity to think further about the social processes that lie behind the façade of dreams.

In terms of cities, it is the quantity of these examples of dreams in waking life that is significant. For me, the occurrence of dreams in the urban landscape demonstrates that dreams are *at least* two-faced: they traffic in both personal and social wishes and anxieties, breaching with ease any barricade that might exist between the personal and the social. Dreams, thus, are important because they are evocative, albeit in a disguised way, of something simultaneously deeply personal and socially significant about city life. Yet, there are so many dreams in cities, they become part of the taken-for-granted world.

Simply listing dreams shows that cities provide a variety of infrastructures for dreams: institutions and places that, we can say, manufacture and house dreams – whether it be in shop windows or railways stations, in arcades or cinemas.[2] If city life is, even in part, made out of a phantasmagoria of dreams; if cities are, even in part, directed towards the production of dreams, then the possibility exists that urbanites are as much dreamt by their cities as they dream within them. What is at stake in the world of dreams is our very grip on the nature of things – and, it must be noted, their grip on us.

Dreams are often used to invoke a better life, a better future. They pull us, in one direction, towards our innermost yearnings and, in another, towards a life beyond the constraints of the real. In an episode (titled *Weightgain 4000*) of the cartoon series *South Park*, Eric Cartman, the fattest of four young boys, decides to tone his body by taking the food supplement Weightgain 4000. Though all Eric does is gain weight, he thinks his body is becoming more and more like that of the buff musclemen that advertise the muscle-building product. On reaching his ideal weight (= huge), Eric echoes the sentiment of the American Dream (see Figure 1.4): 'follow your dreams: you can reach your goals', he implores.

Eric's predicament and his understanding of it is ironic, but it works as irony precisely because of its resonance with the rhetoric of dreams pervasive in western cultures. In talking of dreams, Eric conjures up a yearned-for perfect future world. A world where he

Figure 1.4 Collect the American Dream with Kellogg's.

gets what he wishes for: that is, a real place of wish fulfilment. Alternatively, the place of your dreams can also be somewhere where 'reality checking' is (or has become) impossible or abandoned. In this sense, we become dupes to, or slaves to, our wishes. But to simply suggest that dreams enslave us misses the point (and the irony), for dreams are also expressions of desires and fears that must be taken seriously.

What Eric teaches us is that the word 'dream' (and, welded to it, the idea of the dream) captures people's hearts and minds, explicitly guiding their actions. It is worth remembering, too, that in Eric's exhortation, dreams represent the expression of a wish and the potential for its fulfilment – if you *really* want it. So wrapped up in the dream of the Weightgain 4000 commodity is Cartman that he is unaware that his dreams have not been fulfilled. Indeed, Cartman believes he has achieved his goal. So, Cartman also teaches us a paradox: that fulfilling your dreams is a double-edged sword: be careful what you wish for (a warning we will return to in the next chapter) – especially if it involves the deceitful dream-images produced by (American) commodity capitalism.

For many, dreams are composed primarily of visual images (as in a phantasmagoria). This is not to say that smells, touch, sound and the like, have no place in dreams, but rather that sequences of images are the pre-eminent experience of dreams. This dream-like procession of images is commonplace in the waking world. Indeed, many argue that social life is increasingly mediated by images. For a few, this means that there are direct links between the dream world and the waking world of images (and arguments about the increasing mediation of social life through images). Thus, in *The War of Dreams* (1997), the anthropologist Marc Augé examines what he sees as the pervasive production of a world made up of images. These images, he claims, threaten to take over life itself. For Augé, the danger is that the world will become an occupied country, dominated by images: that is, he says, dominated by dreams. Augé finds evidence that contemporary life is becoming a dream-world of images in, for example, the pervasiveness of television.[3]

Many people have taken dreams to be a code book of the personality. Like a secret agent in a foreign country, the dream would send coded messages to the waking mind. In many bookshops, there are shelves filled with dream books that offer to decipher the secret meaning of your dreams. For many people, interpreting dreams can be a way of gaining guidance for actions in waking life.[4] For example, Mechal Sobel (2000) shows how revolutionary-era Americans used dreams to guide action. Commonly, this works by taking a dream element and its meaning is then decoded in a one-to-one relationship. Dreaming of a sock, thereby, would always indicate an anxiety about poverty. Or, according to another code book, comfort.[5] This method of interpreting dreams – however arbitrary it might appear – remains popular (see Figure 1.5).

Figure 1.5
A small sample of dream books.

Dreams are a guide to life. Indeed, not to fulfil one's dreams can suggest that something has been permanently lost from life. This sense of loss can be exploited. Banks are keen to convince people that they might be able to afford their dreamed-of future life now. You, they coo, can buy

Figure 1.6
Get a 'dream mortgage' in Singapore.

your dreams. Today. When I was in New York in 2004, Washington Mutual Home Loans volunteered: 'Whatever your dream home is, we have your loan'. Similar adverts can be found in Singapore (see Figure 1.6). Nor are British Banks immune. Recently, Lloyds TSB suggested you 'Let Your Dreams Come True' by taking out one of their personal loans. Per capita personal debt, in Britain, is currently at an all time high. Dreamy.

In the business world, dreams aren't just about lending money for future dream purchases. In 2002/03, Honda ran a series of adverts for their cars with the tag line 'The Power of Dreams', in which they made the explicit link between their cars, the power of their imagination and owning a dream car (one of theirs) (Figure 1.7).

Figure 1.7
You can lose (more than) your heart to a 'dream car', London.

Capitalism is more than happy to distil your dreams into a commodity and sell it back to you; or, to ensure that your unformed wishes coalesce around dreams that it already has commodities for.[6] As Franco Moretti put it, 'capitalism is a dream – a bad dream, but a dream nonetheless' (1978a, page 90). Part of the ongoing nightmare of capitalism is that it also wants to harness your dreams to make money. In April 2001, the *Observer* newspaper reported that Helen McLean, a New Zealand psychotherapist, had been employed by Microsoft to help them release the creativity of their workforce through catching dreams.[7]

Meanwhile, IBM were prepared to play on the nightmares of businesspeople to sell its products. Through 2003 and into 2004, they ran TV adverts in which a businessman describes the recurrent nightmares he's having to his psychotherapist.[8] In one of the nightmares, the businessman is fleeing from a chasing pack of people. After some slips of the tongue and word association, the analyst explains to the businessman that the people he is afraid of are his customers. IBM offers the 'perfect business solution' to the nightmare. Yet IBM seems curiously unafraid of presenting the business world as a nightmare and also the (senior) businessman as a paranoid incapable of reflecting upon his own situation. They are clearly persuaded by the dream-like solutions they offer. All businesses, great and small, know that dreams sell (Figure 1.8).

If there is an economics to dreams, then there is also a politics.[9] Dreams have often been used by politicians and activists to suggest a wrecked past and/or a better future. On 9 February 2003, as part of his challenge to Tony Blair's Labour government, the leader of the Conservative opposition, Michael Howard, announced his vision for a caring, conservative

Figure 1.8 'A Great Deal to Dream About'.

Britain. He confidently called this vision 'The British Dream'. Such use of dreams politically speaks simultaneously to personal experience (of wishing for a better future) and to the social experience of injustice and inequality (and, therefore, wishing for a better future). Michael Howard is far from the only person to have used the rhetoric of dreams like this. Most famously, perhaps, Martin Luther King dreamt of a better future:

> Go back to Mississippi, go back to Alabama, go back to South Carolina, go back to Georgia, go back to Louisiana, go back to the slums and ghettos of our northern cities, knowing that somehow this situation can and will be changed. Let us not wallow in the valley of despair, I say to you today, my friends. And so even though we face the difficulties of today and tomorrow, I still have a dream. It is a dream deeply rooted in the American dream. I have a dream that one day this nation will

rise up and live out the true meaning of its creed: We hold these truths to be self-evident that all men are created equal. (Martin Luther King, speech delivered 28 August 1963 at the Lincoln Memorial, Washington DC)

Although, clearly, he was using his personal dream of a better future to link the dream of emancipation to the American Dream, his words connect to the idea that we are somehow free in our dreams. That in our dreams, we are free of the constraints of the waking world. But, it is also clear that dreams are being linked to political activism: dreams are part and parcel of the emotional life of politics – including that within cities (see Figure 1.9).[10]

Figure 1.9 'I Have a Dream' in Sydney, Australia.

If it is assumed that dreams are a space beyond the constraints of ordinary living, then it is also the case that science-fiction stories have mobilised common experiences in dreams to loosen the grip on what is assumed to be real and what is assumed to be dreamt. This idea was effectively used in *The Matrix* (1999) to suggest the reality of a computer-generated dream-world. The plausible reality of this computer-generated world is founded on an analogy with the common experience of dreaming that you've woken up, yet you are still asleep and only dreaming you're awake: how can you tell the difference between the dream world and the real world? Thus, Morpheus explains to Thomas 'Neo' Anderson what the matrix is like:

MORPHEUS: Have you ever had a dream, Neo, that you were so sure was real? What if you were unable to wake from that dream. How would you know

	the difference between the dream world and the real world?
NEO:	This can't be.
MORPHEUS:	Be what? Real?

The last component of the rhetoric of dreams that I would like to tease out is the entanglement of fantasy and desire. As we have heard, Calvino thought that dreams were made of desires and fears. This is a common theme in the phantasmagoria of dreams, one aided and abetted by the sexual liberation allegedly to be found in cities. For example, in the novel by Arthur Schnitzler, *Dream Story* (1926),[11] the protagonist Fridolin, a doctor, ventures into night-time Vienna. He follows a dream-like, nightmarish path through the hidden desires of the city. Eventually, he has to be rescued from danger by a beautiful woman, in a moment both erotic and degrading. Later, he frets that this dream-like experience reveals an authentic desire: no dream can just be dismissed as merely a dream. For her part, Albertine, his wife, is certain that both he and she have survived their adventures and are now fully awake (1926, page 99). Fridolin asks Albertine how sure she is of this:

> 'As sure as I am of my sense that neither the reality of a single night nor even of a person's entire life can be equated with the full truth about his innermost being.'
> 'And no dream,' he sighed quietly, 'is altogether a dream.'
> She took his head in both hands and pillowed it tenderly against her breast. 'Now we're truly awake,' she said, 'at least for a good while.'

Figure 1.10 Dream Girls advertise.

Schnitzler explores the troubled and troubling boundary between the reality of the fantasies in dreams and the reality of acted-out fantasy. Vienna is more than just a setting for the dream story. Vienna itself is dream-like, at least to the extent that it conceals secret desires and fears. Vienna's morality is revealed by Fridolin's erotic (mis)-adventures, showing where the boundaries lie between private dreams and their legitimate public expression. This boundary is a moral one (Figure 1.10): indeed, dreams have been a constant stumbling block for moral and ethical decisions. In this story, the question is whether Fridolin should be held accountable for his secret wishes and desires? Indeed, one might start to wonder whether it is entirely appropriate to contrast reality and dreams. And can it be good or right to side with reality against dreams?

Dreams are part and parcel of the commonplace experience of everyday city life: in shop windows, in cinemas, on television, on bookshelves, in ordinary understandings about what it means to lead a better life, in moral codes about appropriate behaviour, in the ways we learn to see the surface appearance of urban landscapes, and even as something experienced during sleep. Dream phantasmagorias are operative and effective in many different places and under many different circumstances. Indeed, they are almost irrepressible: perhaps this is because dreams tap into secret desires and fears, as Calvino suggests, or maybe because they enable the expression of feelings of yearning and anxiety (that so often are associated with hidden desires and fears), both personal and collective. In this light, it is useful, rather than confusing, that the word 'dream' refers simultaneously to personal (sleeping) and collective (waking) experiences.

In these instances of dreams, we can discern that there is a certain intensity of emotion associated with the phantasmagoria of dreams. By using dreams, politicians and banks, writers and activists, can evoke particular emotional qualities of life, including city life. Dreams appear able to weld personal and social, sleeping and waking, experiences together – or, at least, to create a bridge, across which these experiences can move. I have suggested, also, that through dreams there is an emotional traffic, often involving desires and fears. This traffic is not always comfortable. Indeed, the results of these movements of thoughts and feelings through dreams can be quite disturbing. Often, the solution appears to be to wake up or to get real. But waking up or getting real isn't that straightforward, as dreams are woven into the fabric of the real world. To show this, I will look at the ways in which cities appear in two stories about dream-worlds. The consequences of waking up can send shivers down the spine.

1.3 Cities of the Imagination

There can be something reassuring about the phantasmagoria of dreams in waking life. It is as if something quite strange within us – our dreams – has been given the freedom to be. As Lefebvre observes, there is something *simultaneously* foreign *and* intimate about dreams: 'the space of the dream is strange and alien, yet at the same time as close to us as is possible' (1974, page 209). More than this, dreams liberate desires: the desire for commodities (such as cars, mortgages and the like); the desire for freedom; the desire, indeed, for desire. There's something else. Dreaming can create worlds that are both strange and familiar: uncannily familiar worlds that are nonetheless strange and disturbing. To show you what I mean, I'd like to talk about a couple of stories that appear in Neil Gaiman's graphic novel series, *The Sandman* (in 75 parts, 1988–96). In both these tales, cities figure prominently. These cities are both strange and familiar, but what is really scary is what happens when the dreamer wakes up. Describing this uncanny affect, this creepy side of dreams and cities, is the focus of this section.[12]

Neil Gaiman is a significant figure in the world of graphic novels.[13] Gaiman's first story was an episode of *The Swamp Thing*; a series given new life by Alan Moore, also a significant writer of comic books. It is worth pointing out that the work of these two writers overlaps in other ways. For example, in one plotline for *The Swamp Thing* Alan Moore created the character John Constantine, a dodgy English magician ('Growth Patterns', Moore, 1985). Later, Neil Gaiman would write a John Constantine story himself ('Hold Me', 1995) and he is also a central figure in another Gaiman comic, *The Books of Magic* (1990–91).[14] John Constantine's popularity – based in part on the horror of the stories and in part on the intense ambiguity of the character – led to him being given his own graphic novel series, *Hellblazer* (1987 onwards).[15] Indeed, the intertextuality of these stories in itself has dream-like qualities and effects, creating overlapping worlds of fantasy and imagination.

Gaiman's major work is undoubtedly *The Sandman* series. The central figure in this series is Lord Morpheus, who is King of the Realm of Dreams. There is much that can be said about Morpheus and his tragic fate.[16] However, my interest is in two stories about the interweaving of dream-worlds and cities. One deals with the irony of the preservation of a real city in the world of dreams, another with the horrifying possibility that the city might be dreaming its inhabitants. These stories can also tell us much about the grip that dreams have on us and what is at stake, emotionally, in thinking that dreams blur the edges between the sleeping and waking worlds.[17]

'Ramadan' (1993a) tells the story of Haroun Al Raschid, King of Kings, Prince of the Faithful who rules in Baghdad, the Heavenly City. The graphics show Baghdad as a fabulous city – full of Oriental splendour. In the tale, Baghdad is described as the most perfect of all cities:

> There was no court that was like to Haroun Al Raschid's. He had gathered to him all manner of great men from all corners of the world. There were sages and wise men, and alchemists, geographers, and geomancers, mathematicians and astronomers, translators and archivists, jurists, grammarians, cadis and scribes. In his court were the greatest teachers of the Hebrews, who were the first of the three people of the book and the greatest monks of the pale Christians (a dirty folk, who will not bathe, and who venerate the dried dung of their leader, whom they call the Pope) … and, as you must realize, he had with him the greatest scholars of the Kuran, the word of Allah, as revealed to his prophet Mohammed, one hundred and eighty years before. (page 228)

It was a city of pleasure, wealth, myth, of wonders both technological and organic. A cosmopolitan city that housed everything and everyone, 'but Haroun Al Raschid was troubled in his soul' (page 232). Eventually, the Caliph of Baghdad succumbs to his troubled thoughts and he forces the Dream King, Lord Morpheus, to appear before him. Morpheus is furious and forbidding. Undeterred, the Caliph wishes to make a bargain. He uses a magic carpet to show the Dream Lord his city and its great marvels and commerce. Morpheus is quite impressed. Then the Caliph asks 'how long can it last? How long will people remember?' (page 253). Haroun Al Raschid predicts that Baghdad will eventually turn to ruins and become covered over by the sands. It will disappear and become forgotten. 'This is as good as it's going to be, isn't it?' observes the Caliph, melancholically (page 254).

The Caliph's solution is simple: that Morpheus takes the Heavenly City into the realm of dreams, where it will last forever. Morpheus replies: 'All you need do is tell your people. They follow you, after all. And yours is the dream' (page 254). The deal is done. Here is the twist in the tale. The Caliph wakes up on a dirty mat, around him the city of Baghdad is in ruins. He half remembers a dream, but it is already fading. Haroun Al Raschid spots a tall stranger in the market. The man is holding a glass jar and in the jar is a golden city. The Caliph is hypnotised by the city under the glass, but does not recognise it. The splendid, heavenly city is preserved forever as a dream, but barely remembered on awakening.

In the *Worlds' End* sequence of tales (Volume VIII), a group of travellers are forced by a fierce storm to seek refuge in an inn. To pass the time while the storm rages, the refugees take turns to tell tales. Mister Gaheris begins his. It is 'A Tale of Two Cities' (1993b): 'There once was a man who lived in a city, and he lived in that city all his life' (page 27). The man, Robert, lived an ordinary life. He worked, commuted, read newspapers, went on holiday. And he wondered occasionally about the world around him. But he also had a passion for observing the city. Like the best psychogeographers, he took note of everything:

> Robert saw the city as a huge jewel, and the tiny moments of reality he found in his lunch-hours as facets, cut and glittering, of the whole. Is there any person in the world who does not dream? Who does not contain within them worlds unimagined? (page 28)

At one point, Robert catches sight of a Silver Road. He runs towards it, but it simply vanishes. All day, Robert is disturbed by this and he day-dreams about the Silver Road. Later, when Robert takes the train home, he finds he is entirely alone. Except, that is, for a mysterious stranger – whom readers would be able to identify as Morpheus. Robert's anxiety is now even greater and he wonders whether he is on the right train. He stumbles out at the next stop, but he doesn't recognise the station. Even so, Robert feels sure he knows the city well enough to be able to orient himself and grab a cab home. As he looks around, there is something familiar about the street but he is unable to place it or to name it. Robert hurries through the streets, certain that he will eventually come to a place he knows. The sights and the smells of the city are so familiar, yet so strange. In panic, he runs through the streets. Eventually, Robert collapses exhausted. He is sure there are people around him, but they appear to be like ghosts. They shimmer and vanish. No one approaches him or even passes close by. Now, as Robert walks,

> The roads mixed him up, turned him around. Here, he would pass a cathedral or museum, there, a skyscraper or a fountain – always hauntingly familiar. But he never passed the same landmark twice, could never find the road to return him to the landmark again. (page 33)

He is not exactly lost because the city seems so familiar, yet he is also totally lost. He walks for hours, maybe months. Eventually, Robert meets an old, ragged stranger on a bridge. 'It's beautiful, isn't it', the man says, pointing to the city. Robert wants to know where in the city they are. Cryptically, the stranger answers.

Perhaps the city is a living thing. Each city has its own personality, after all. Los Angeles is not Vienna. London is not Moscow. Chicago is not Paris. Each city is a collection of lives and buildings and it has its own personality. (page 35: see Figure 1.11)

'So?' asks Robert. The old man replies,

So, if a city has a personality, maybe it also has a soul. Maybe it dreams. That is where I believe we have come. We are in the dreams of the city. That's why certain places hover on the brink of recognition; why we almost know where we are. [...] I mean that the city is asleep. And that we are stumbling through the city's dream. (page 35: see Figure 1.11)

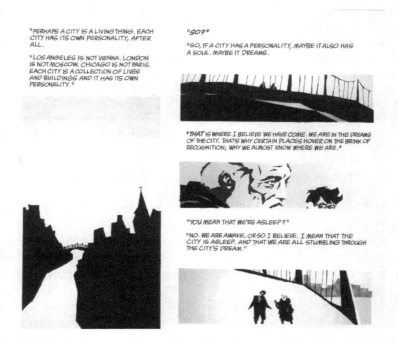

Figure 1.11 From Neil Gaiman's 'A Tale of Two Cities'. Reproduced by kind permission of DC Comics, Inc.

The ghostly figures, argues the stranger, are simply those people who briefly enter the city's dream before leaving it. Or maybe, he speculates, they are simply temporary figures in the city's dream. Robert begins to wonder what will happen to him. Will he, for example, return to the waking world? Could he find a path back to the real city? Never mind that, the old man laughs, what if the city wakes up?!

But now Robert had a purpose, he looked for something he knew: a path, or street or alley; he walked the city of dreams hunting for something he recognized: searching for the real. (page 37: see Figure 1.12)

Figure 1.12 From Neil Gaiman's 'A Tale of Two Cities'. Reproduced by kind permission of DC Comics, Inc.

Close to death,[18] Robert makes one last headlong dash towards a doorway that has just appeared before him. Once through, Robert is stunned to find himself back in the Real City. And so Mister Gaheris ends Robert's story, but he hasn't quite finished his tale. Mister Gaheris then informs the attentive storm-blown travellers that he had once met Robert in a small hamlet in Scotland. Gaheris had asked Robert whether he feared the city of dreams, was this why Robert was now living in the wilderness? Robert replied:

> If the city was dreaming [...] then the city is asleep. And I do not fear cities sleeping, stretched out unconscious around their rivers and estuaries, like cats in the moonlight. Sleeping cities are tame and harmless things. What I fear [...] is that one day the cities will waken. That one day the cities will rise. (page 40)

What if? In both the tales, Gaiman is seeking to produce an uncanny effect: the shivers are meant to creep up and down the spine; the hair to stand up on the back of the neck. In the first story, the Caliph wishes to preserve his ideal, dream city, but he can only do so in dreams. On waking, the dream city is lost to the waking world – even as it is preserved forever in the dream. This idea is reminiscent of Freud's suggestion that the unconscious has the ability to preserve experiences, yet this risks the possibility that material placed in the unconscious might return unexpectedly, undesirably.[19] The dream-world, in this sense, acts as a defence both against losing things and also against waking up to a frightening real. But, uncannily, the Caliph has woken up to a strangely familiar real, in which he has lost everything. The horror is not just that the Caliph has lost the very thing he sought to preserve, but also that he now lives in an impoverished Baghdad: he is in rags, the city in ruins.

In the second tale, Robert realises that he is in a familiar place, yet it is also strange; a strange place, that is also curiously familiar. He is increasingly disconcerted by this experience, as the uncanniness of the experience dawns upon him. Here, Gaiman is evoking the strange, but familiar, experience of getting lost in the city. Indeed, in his essay on the uncanny, Freud's own reminiscence of an uncanny experience involves him getting lost in the city, and finding himself – against his will – back in the same place.[20] What Robert's experience demonstrates is how dream-like these uncanny experiences (of the city) are, but also how familiar and strange ordinary experiences (of the city) can be. In many ways, the city provides an archetypal scene for uncanny experiences; experiences that, as Simmel's work would suggest, city dwellers seek to protect themselves against, perhaps through a blasé attitude or indifference.[21] This suggests, further, that emotional responses to dream-like and/or uncanny experiences of the city off-set, or shift, their emotional intensities in a variety of ways.

In both of Gaiman's tales (as well as Freud's own experience of getting lost), the twist is directly related to the express wishes of the dreamers and how these are thwarted. These are tales of the unexpected, and ironic, fulfilment of wishes. It's the old caveat: be careful what you wish for – for you might just get it. Robert, for example, wishes desperately to awake from the dream, but then worries that the consequence of awakening is that he will cease to exist. But this is not the end of matters for Robert. On waking, he becomes afraid that his wish to wake up will return to take vengeance upon him – if and when cities themselves wake up.

Robert's ultimate fear is that cities will wake up, like a deadly predator, from a deep slumber to tyrannise their human inhabitants. In this image of the sleeping city, Gaiman poses the incredible possibility that cities dream their inhabitants (and not vice versa); that urban dwellers, that city life, exists only in so far as it is dreamt by a sleeping city. This is an intriguing idea, for it suggests that city dwellers live within the dreams that cities create for them (alternatively, see Figure 1.13). This is not such an incredible suggestion given the evidence presented already. However, we can now take this suggestion one step further.

The implication is that the work of dreaming that goes into making dreams is just like the work of dreaming that goes into making city life and into the production of urban space. On this basis, we can wonder whether the hidden work that goes into making the city's phantasmagorias,

Figure 1.13 A mural depicting the city in Dreamtime (Sydney, Australia).

its dream-worlds, is also like dream-work. To explore this question, the next logical step is to look at how dreams are made. Therefore, I will look carefully at how Freud interpreted dreams (in section 1.4) and then extrapolate how this model of dream-work might aid our understanding of, what I call, city-work (in section 1.5).

1.4 Dream Analysis and Dream-Work

The rhetoric of dreams has been applied, liberally, to cities and to aspects of cities. There are many examples. Most great cities of the world have been described as 'dream cities' by someone: Le Corbusier on New York, most people on Paris, and so on. Similarly, city planners' and architects' designs are often described as 'dreams', as are film-makers views of cities.[22] Stories about dreams, moreover, contain references to cities — whether in the work of Neil Gaiman or G. K. Chesterton (see above and note 17). There seems to be a commonplace desire to weave together the web of dreams and the fabric of cities.

It is also the case that, often enough, in a variety of ways, cities turn up in dreams. One such dream, as analysed by Sigmund Freud, will provide the example for this section. Freud's interpretation of dreams, meanwhile, provides us with four key ways to think about the work that goes into making dreams. These are condensation, displacement, representation and secondary revision (1900).[23] I will argue, in this section and the next, that the production of dreams is much like the production of space. I will show that dreams are produced *spatially* through such means as condensation and displacement. This model of dreams then acts as a way of both understanding the significance of dreams in city life as well as allowing the more fantastic and imaginative aspects of city life to become central to any analysis of the real of city life. And it is helpful to remember, at this point, that Freud's interpretation of dreams interferes with a strict dichotomy between internal and external worlds, be they personal or social.

Freud's basic (and still contested) insight is that dreams are meaningful. He forcefully argues throughout *The Interpretation of Dreams* that it is not possible simply to decode dreams as if their symbols always and inevitably refer to one thing: for example, as if dreaming of being in rough seas always means insecurity or vivid colours represent a turning point in one's life. Instead, Freud argues that disclosing the motivating thoughts of the dream requires tracking back through the devious and intricate operations that the mind has performed in order to construct the dream. It is the sheer deviousness and intricacy of dreams that is interesting, because it suggests a world of meaning and significance that is not simply decodable or manipulable. There is an ever-present openness, flexibility, interconnectedness and cunning in dream-worlds: dream-worlds *both sleeping and waking*. Later, we will explore how dream-analysis might help interpret the waking world, but for now let us focus on Freud's account of dream-production.

It is hard to overestimate how important Freud considered dreams, both for psychotherapy and psychoanalysis.[24] Their significance is often attributed to Freud's belief that dreams provided a royal road to understanding what was happening in his patients' unconscious minds. Elsewhere Freud would also argue that dreams themselves form the prototype for a wide variety of psychic systems of thought, whether personal (such as paranoia, neurosis and psychosis) or social (such as magic, animism and religion).[25] However, I am more interested in how Freud conceptualises dreams spatially. The promise

Figure 1.14
Brion Gysin and Ian Sommerville's *Dreamachine* (1960) was designed to stimulate visual experience and image formation and to disseminate new forms of consciousness (see Hiller, 2000). A cardboard cylinder revolves at 78rpm, while the light bulb projects light outwards. The viewer looks at the device with their eyes closed, experiencing the flickering of light and dark and of kaleidoscopic colour on their eyelids. Gradually, the patterns experienced become ever more complex and ever more beautiful, as in a dream.

is that the way in which Freud thought *spatially* about the dreaming mind[26] can provide a model for thinking about spatialities more generally — and specifically the production of urban space.

It is important to recognise that Freud argues that dreams do not arrive in the mind fully formed. *Dreams have to be made* (Figure 1.14). For Freud, the material used in the construction of dreams can be drawn from every aspect of experience: from childhood experiences and impressions to something that has just happened; from pure fantasies to sensations emerging within the body. These memories, experiences, impressions, sensations, fantasies, feelings, situations, places — whether conscious or unconscious — provide the raw materials, but they will be selectively deployed in the construction of a dream. The creation of a dream is motivated by a dream thought, or set of dream thoughts. These thoughts are the latent content of the dream.

Latent content is implicit within the dream: there is no direct, or directly decodable, access to these thoughts. This is because dream-thoughts contain material that is too troubling to be expressed directly. These disturbing thoughts can have many different motivations, but they form into a wish of some kind: a wish that cannot be directly expressed, as it would shock the dreamer and wake them up. Freud's famous formula for dreams is that they are '*(disguised) fulfilment of a (suppressed or repressed) wish*' (1900, page 244).[27] Dreams are therefore also '*the GUARDIANS of sleep*' (page 330).[28] In the production of the dream, the raw materials, the latent dream-thoughts and their emergent wishes, are covertly worked upon (many times over) to create the *manifest content* of the dream: that is, the (seemingly indecipherable) dream-images as recalled and related by the dreamer. This deceitful manufacturing of dream-images is known as dream-work.

Freud identifies four basic forms of **dream-work**. For Freud, the most important mechanisms are condensation and displacement. These are aided and abetted both by the means of representation (in the dream) and by secondary revision (as the dream is recalled and related). My main interest is in the fact that condensation and displacement are both spatial operations: about convergence on, and divergence from, specific elements in the dream. This is suggestive for thinking about how fragments of urban space might draw 'things' (images, feelings, meanings, etc.) towards them, or push them away. How dreams represent one thing for another, and draw upon familiar and unfamiliar means to do so, is suggestive of the relationship between the manifest and latent content of urban landscapes (as in a phantasmagoria). Meanwhile, secondary revision becomes important in thinking about urban space because it both creates a narrative for space (as in a dream) and also gives space a façade – a mask that simultaneously conceals and expresses its own production (as in a phantasmagoria). To exemplify dreamwork, I will use Freud's analysis of 'A Lovely Dream'.

I have chosen 'A Lovely Dream' both because it is set in a city (in analysis, revealed as Vienna) and also because Freud discusses condensation and displacement in terms of it. The patient's dream (as recorded by Freud) goes like this:

> He is driving with a large party to X Street, in which there was an unpretentious inn. (This is not the case.) There was a play being acted inside it. At one moment he was the audience, at another actor. When it was over, they had to change their clothes so as to get back to town. Some of the company were shown into rooms on the ground floor and others into rooms on the first floor. Then a dispute broke out. The ones up above were angry because the ones down below were not ready, and they could not come downstairs. His brother was up above and he was down below and he was angry with his brother because they were so much pressed. (This part was obscure.) Moreover, it had been decided and arranged even when they first arrived who was to be up above and who was to be down below. Then he was walking by himself with such difficulty and so laboriously that he seemed glued to the spot. An elderly gentleman came up to him and began abusing the King of Italy. At the top of the rise he was able to walk much more freely. (1900, page 390)

A measure of the sheer quantity of thought that goes into even this short dream is that Freud's analysis of the dream is over seven times longer than the dream itself (see pages 390–395, 414 and 439–40). The interpretation of dreams, for Freud, begins by taking individual images or fragments of the dream and asking the dreamer to consider what these images or fragments remind them of (sometimes using a technique known as 'free association'[29]). Through these associations, Freud seeks to track the meanings and feelings from which the dream was formed. Thus, in analysis of 'A Lovely Dream', the patient reveals an association between the difficulty of walking in the dream and the memory of breaking up with a girlfriend. The patient is able to make the connection between walking and the break-up through an intermediary association, a poem by Sappho. The break-up had been painful: had the dreamer realised this was the meaning contained in his difficulty walking in the dream, he was sure to wake up. In this case, the dream is also hopeful: the dreamer's burden gradually lightens and, by the end, he is able to walk freely. Each element, like the difficulty in walking, in the dream can be 'unbundled' in this way.

For Freud, each element – or dream-symbol – not only refers to many meanings, but also permits many meanings to exist in the same space, however contradictory they might seem:

> Not only are the elements of the dream determined by the dream-thoughts many times over, but the individual dream-thoughts are represented in the dream by several elements. Associative paths lead from one element of the dream to several dream-thoughts, and from one dream-thought to several elements of the dream. Thus a dream is not constructed by each individual dream-thought, or group of

dream-thoughts [but] rather, by the whole mass of dream-thoughts being submitted to a sort of manipulative process in which those elements which have the most numerous and strongest supports acquire the right of entry into the dream-content. (1900, page 389)

Freud argues that one dream-element can represent many different dream-thoughts (and vice versa). Many dream-thoughts (ideas, fantasies, experiences, memories, sensations, etc.) can be combined — condensed — into one symbol that simultaneously refers to them all. *Condensation* is Freud's term for the work that goes into making the 'nodal points', at which many trains of thought intersect, into dream-symbols. Significantly, the more a symbol can act as a *point of convergence* for many trains of thought, the more likely it is to be admitted to the dream (1900, page 388, also 456–457). Dreams, therefore, are likely to be populated by dream-elements that are saturated with meaning:

[...] the elements formed into the dream are drawn from the entire mass of dream-thoughts and each one of those elements is shown to have been determined many times over in relation to the dream-thoughts. (1900, page 389)

Let us return to one element in the above dream: the difficulty in walking. Through analysis, it becomes clear that this element refers simultaneously to difficulty in breathing once experienced by the patient, to the inhibition of movement in response to an anxiety, and to a cautionary poem by Alphonse Daudet. One element determined three-times over. Or, one element with three trains of thought running out from it. By following the trains of thought, it is possible to track the links of association further and further. Daudet's poem, indeed, allows Freud to make links to the patient's love-affair with an actress, who lives in X Street. In this way, Freud can show that dream-thoughts can be represented more than once, and that trains of thought can intersect and interact with one another.

In analysis, moreover, certain elements can set in motion further trains of thought. Thus, the inn reminds the patient not only of where he met his lover, but also of a conversation with a cabby at the time. In this conversation a further connection is made to the inn not being a proper hotel. This then leads the dreamer to make yet another link, between the inn and vermin: a particular phobia of the patient. From there, the patient remembers yet another poem and a further train of thought is opened up. This one leads to an association between the actress and his wet-nurse and this, thereby, opens up a meaningful association between the dream-symbol 'inn' and 'lovely breasts' (as the analysand puts it). And so on. Dreams, then, can be related to ongoing phobias and inhibitions (1900, page 605)[30] as well as the re-emergence of anxieties and desires in past experiences, especially from childhood (for example, pages 696–697).[31]

Condensation works to admit dream-thoughts to the dream — and to avoid the censorship of dream-thoughts — by using seemingly disconnected symbols. It achieves this using a vast array of techniques: through association, collaterals, combination, composition, substitution, surrogation, congruence, proximity, 'just like' logical relations, and reversal into opposites. It is important, also, to bear in mind the work that goes into selecting the spatial arrangement and the spatial setting of the dream. Thus, the dreamer places his brother in the rooms *above* him as this puts the brother in a socially superior position to himself; and, the dream is set in a street, in an inn and in a theatre because these have associations both with particular events and with social standing:

> Dreams construct a *situation* out of these images; they represent an event which is actually happening [...] they 'dramatize' an idea. (1900, page 114)

Thus, the dream sets up a situation – for example, a large party, with a play inside it – which will eventually end in the dreamer walking home alone. *At the same time*, a spatial setting is constructed that promotes the dramatisation of the dream's concealed wish, albeit in a deceptive way.

Displacement is closely allied to the work of condensation (and uses its operations), yet its role is to ensure that the dream is '*differently centred*' from the (dense, intense) emotional core of its dream-thoughts (1900, page 414). That is, displacement ensures that the crux of the dream appears to be somewhere else from where it really is. For example, in 'A Lovely Dream', Freud argues that:

> the central position was occupied by *climbing up* and *down* and being *up above* and *down below*; the dream-thoughts, however, dealt with the dangers of sexual relations with people of an *inferior* social class. So that only a single element of the dream-thoughts seems to have found its way into the dream-content, though that element was expanded to a disproportionate extent. (1900, page 414)

Displacement's purpose, therefore, is to misdirect the dreamer and it does so by shifting where the emotional centre of the dream seems to lie and also by disguising the emotions involved. This implies that there is a *transvaluation* in the psychic value and intensity of the dream, so that the (emotion of the) wish that lies behind the dream is placed somewhere away from the centre of the dream itself. This transvaluation involves placing the latent psychic intensity of a thought at a remove. This is often achieved by using situations that are reminiscent of the emotion of the dream-thought, but that do not lead directly back to the dream-thought itself. This effectively disguises the emotional content wish of the dream while paradoxically producing the feeling that the wish has nonetheless been fulfilled.

In many ways, displacement works by *divergence*: either by off-centring the dream away from its central affect, such that the affect stays in its proper place but the focus of the dream seems to be elsewhere; or by off-centring the affect in the dream, pushing the emotional centre of the dream on to seemingly marginal, or irrelevant, dream-symbols.[32] Most commonly, displacement uses 'reversal into opposites' (where an emotion is 'represented' by its opposite and thereby disconnected from it), but it can also work through using parallels (where the emotion used is just like, but not, another), subsumption (of one intensity into another) and diminution/exaggeration (making the emotion have a lesser/greater intensity than warranted).[33]

In 'A Lovely Dream', Freud argues that the emotional core of the dream circulates around the patient's anxieties about having sex with women of a lower social class. Displacement ensures that this anxiety is disguised (and paradoxically expressed) by a series of (simultaneously social and spatial) settings in which substitute people are placed above or below one another, representing higher and lower social status. The higher and lower social positionings persist in the dream (though they could just as easily have been inverted), while the dreamer's anxieties about sex, desire and women's bodies have become both remote and suppressed. In this light, it is unsurprising that the dream ends with the dreamer walking freely on reaching the top of a rise (and that a *rise* should be chosen for the expression of freedom).

For Freud, condensation and displacement were the two overseers of the production of dreams (1900, page 417). Together, they suggested that understanding dream-elements would require following the myriad pathways to and from each element. The third form of dream-work that Freud considered significant concerns *issues of representability*. Issues of representability (as with condensation) revolve around the necessity of evading the censorship of the wish contained in the dream (1900, page 459–460). Due to censorship, restrictions apply to the dreamer's selection of dream-symbols. Moreover, not every dream-symbol will be adequate to the sheer quantity of dream-thoughts that it may have to represent. Dream-work, therefore, also involves the creation of adequate dream-symbols. It is often the case that no adequate 'off-the-shelf' dream-symbol is available and that a 'composite' dream-symbol has to be made up out of several possible alternatives.

The selection or creation of dream-symbols can have unexpected consequences. Thus, a dream-symbol selected to represent one dream-thought might set in motion other trains of thought. In part, issues of representability arise out of the compromises inherent in dream-production. It is in the logic of dreams that they wish to represent all dream-thoughts in some way, even when confronted by logical inconsistencies or incompatible choices (Figure 1.15). Thus, both sides of a contradiction or opposition can easily be represented in the dream; often by the reversal of opposites, but also by the use of ambiguities (that contain both possibilities).

Figure 1.15 David Robillard's 'The Yes No Quality of Dreams' (1988) evokes the play required by the dream's inability to choose. The painting was selected for the Dream Machines exhibition in London, 2000 (see Hiller, 2000).

The fourth and final aspect of dream-work identified by Freud is called *secondary revision*. Freud notices that

> not everything contained in the dream comes from the dream-thoughts, but rather that a function of the psyche indistinguishable from our waking thoughts can make some contribution to the dream-content. (1990, page 319)

The dream is given a working-over to make it intelligible, while dreaming, and also after waking. Secondary revision fills in the gaps in the dream structure, making it more coherent and less absurd. An intelligible pattern is woven out of the dream elements by adding further elements. Freud suggests that a façade is built on to the structure of the dream, but only where there is material available that fits the forming dream.

Freud's analysis of dream-work specifies the means through which dream-thoughts are converted into dream-images. It suggests that conflicts over the expression of dream-thoughts and wishes *within the minds of dreamers* cause dreamers to go to great lengths into fooling themselves that the dream is both real (while asleep) and absurd (on waking). This ability of the dreamer to create a deceitful dream-world is worth pondering over. The dreamer, for a while, is entirely capable of producing a dream whose reality is sufficiently believable that the dream may continue, however bizarre that world seems on waking. Who is to say that the waking world is not full of such deceptions? The material presented in section 1.2 might alert us to the idea that the waking world might also be equally absurd and bizarre, yet we moderns have not woken up to this reality.

The psychoanalysis of dreams, at least, offers the possibility of interpreting the phantasmagoria of dreams in waking life. The symbol of the dream acts as a point of intersection for many trains of thought and is determined many times over with meaning; affect is displaced on to or away from the symbol of the dream, while issues of representability and the need for a coherent narrative are solved by bringing 'dreams' into the waking dream-world. Much the same can be said of other elements, other phantasmagorias, found in cities. In this light, the work that goes into making the phantasmagorias of modern city life is analogous to dream-work. Each fragment of the city, thus, would have to be treated as a deceit to be situated in networks of affect and meaning. Tracking the chains of association that run out from each element might offer a way to discern the hidden, disturbing wishes that go into making up cities and city life. Dream-work, in other words, might be fundamental to understanding the work that goes into producing the phantasmagoric spaces of modern city life.

1.5 Dream-work and the Work of City Space

Successful dream-work depends on making undisturbing worlds plausible. It draws on the mass of available images in the world to create a setting for the dream, a setting which enables the dream to unfold. To create these worlds, dreams are formed out of ordinary geographies, such as sites, journeys, places and the like (Freud, 1900, pages 435–436). These dream-spaces can be *composites* made up by combining different geographical spaces (1900, pages 431–437). The significance of dream-spaces, as with other dream-images, is that they refer to many possible dream-thoughts: just as the inn does in 'A Lovely Dream' (see above). Similarly, directions, such as going up or down, or placing things side by side can be used to imply connections, or relations, between things. I have far from exhausted the possibilities here. What is important is that *spatial relations* – such as setting, sequencing, juxtaposition, reversal, convergence or divergence, distribution, procession, movement, motion, proximity and distance, absence and presence, direction, architecture, comportment, combination and composition – *are not just a passive backdrop for the dream, but part of its very construction.*

It would be quite easy to make a one-to-one correlation between this spatial production of dream-worlds with the dreaming production of urban space. Thus, cities are made out of locations, journeys and localities that condense meanings in a variety of ways. Squares, parks, skyscrapers, for example, are all determined many times over: by affect, by meaning, by social relations. Often, the designs for building seem to be 'composites' of previous designs, half remembered from other times. Such spaces are the waking dreams of architects, developers, owners and/or planners made into grass and tarmac, glass and steel. Not that it stops there. Fragments of cities attract, or deny, feelings that are often inappropriately intense. And every street sign and traffic direction suggests some relation between one place and another, one history and another. And so on. Thus, city life is spatially constructed out of the mass of raw materials available to it – deliberately, thoughtfully and also deceitfully so. However, this is to treat the city as if it were *a* dream; city life as if it were *only one* phantasmagoria. I am more concerned with the ways dream-work illuminates (what I call) **city-work**: the work that goes into making any number of urban phantasmagorias, as real experiences of city life. In this section, I will describe the valuable lessons dream-interpretation can teach an analysis of the real life of modern cities and their phantasmagorias. There are four important bundles of ideas to get hold of.

First, Freud's understanding of how particular symbols in dreams are selected or created is helpful in thinking about the **emotional work** that goes on in cities. Freud's argument that dream-symbols are overdetermined suggests that elements in the city are points of intersection for many social processes and are determined many times over. Similarly, Freud shows that each dream is produced such that its emotional core is not where it appears to be. Following this, it can be suggested that emotions in cities – their intensity, their density, their locations – are commonly displaced; displaced on to marginal social figures and spaces or by shifting the emotionally intense aspects of cities somewhere other than they should be. This may mean many things. For example, it could mean that 'power' (social relations of power) and 'affect' (social relations of emotions) do not manifest themselves directly. Or it could mean that what is important in city life isn't quite where you think it is. Such an appreciation of the mobile networks of affect, meaning and power in cities implies this: in order to get at some of the real (really operative) processes in city life, attention should be paid to those things that appear marginal, or discarded, or lost, or that have disappeared or are in the process of disappearance. Like dream-images that slip through the fingers on waking. It also suggests this: particularly distracting or intense places within cities are part and parcel of the production of the phantasmagorias that cast a veil over city life.

Dream-work continually works material over and over. In doing so, it renders the latent content of the dream implicit, just out of reach. In particular, causality is disguised. Sometimes this is achieved by making closely related experiences far apart or by altering the spatial connections between things so that the causal links between them are reversed (for example, by putting someone of lower social standing on a pedestal). Causality in this case is determined spatially, but there are also significant issues around time: causality is often implicit in chronology, time flowing from cause to effect. Dreams can manipulate time as easily as space: the temporal architecture of dreams also can tell us much about the co-existence of different times in cities. The second aspect of dream-work I'd like to emphasise concerns its **time-work**, especially the use of 'non-linear' times and of the 'preservation' of time. Dreams mix times in specific ways: chronologies are reversed in order to disguise causality, the present modifies the past, events from different periods are aligned, or simultaneous events separated, and so on. These temporalities make time plastic, manipulable. This is especially useful in interpreting those aspects of cities that appear fixed, or long-standing, or that disrupt time, or disappear, or belong to different times: examples we will encounter later include monuments and ghosts.

Third, the spatial architecture – or **space-work** – of dreams is fundamental to dream-work (the production of dreams). This means that in dream-interpretation, or analogously in city-interpretation, it is important to look at how those spaces are made. Space has long been held to be fluid, malleable, non-linear. That is, space is commonly understood to be produced and changed by social processes. It is important, therefore, to understand that these social productions of space are free neither of dream-like qualities nor of the spatial operations fundamental to dream-work, such as setting, sequencing, juxtaposition, reversal, convergence or divergence and so on. Dreams are not simply personal affairs, locked and isolated within the dreaming mind. Dreams are part of the dream-worlds of city life. It is not such a far step, then, to suggest that the space-work of dreams is similar to the space-work of city life: through deliberate interventions to produce particular kinds of space. This will work in as many ways as space-work does in dreams: that is, by spatial operations like procession, direction, distribution, movement, motion, proximity and distance, absence and presence, compositions and so on. Appreciating this will be especially useful for interpreting those elements in cities that *travel* in various ways, particularly where the affect, meaning and power associated with the element changes as it moves: an example that I will develop is the vampire, which inhabits and circulates through cities, changing as it does so.

Fourth, Freud boldly states that dreams are the disguised fulfilment of repressed wishes. This simple formula is far from simple in its ramifications – and the discussions of dream-work and space-work stem from this complexity. Nevertheless, though disguised, the wishes behind dreams appear to be relatively simple: on the one hand, there are repressed wishes located in the unconscious (perhaps the concealed desires and fears Calvino hinted at above); and, on the other hand, there's the wish to remain asleep. Wishes, moreover, are intimately connected to anxieties, inhibitions and the like: with dreams, there are also nightmares. Once we begin to think about cities, though, we must take into account **the sheer variety of desires and anxieties** that might motivate space-work. Even so, dreams teach us that neither desires nor anxieties are directly expressed. Indeed, they are likely to be represented through devious and deceitful images, through ambiguous images, or even through reversals into opposites. If it were otherwise, city dwellers might be shocked into recognising their wishes for what they are: that is, they might wake up to their dreams.

It must remain in question as to whether city dwellers really wish to awaken from, or to, their dreams. That is, whether they'd rather sleepwalk through the modern city, remain distracted and entertained by the city's phantasmagorias (whether in the shopping malls or the cinemas), or, alternatively, whether they wish the city itself to remain forever asleep – as Robert did in 'A Tale of Two Cities' (see above). At this point, we are close to certain arguments Walter Benjamin makes about the relationship between dreaming, cities and modernity.

The work of Walter Benjamin is the final step to take in this exploration of dreams and cities, partly because of the weight he gives to the role of dreams in city life. In the following two sections, I will examine Benjamin's ascription of dream-like qualities to the city and city life. Benjamin's dream-analysis of city life reaps several rewards. It allows us to identify the phantasmagorias of modern city life, to see the work that goes into making the dream-like qualities of city life, and to track the variety of meanings that can attach to elements and images in the city. This dream-analysis will usefully begin, in the next section, with Walter Benjamin's identification of the dream-spaces of modernity, as this is where the dream-work in the production of the modern city is most apparent.

1.6 City Space and the Web of Dreams: Paris and London

Most commentary on Walter Benjamin's writings on modernity and the city has noted his use of dreams.[34] Freud's model of dreams afforded Benjamin with a way to both describe and analyse the production of city space. Benjamin himself assiduously collected the dream elements of the city, for later analysis. Thus, for example, Convolute K of his *Arcades Project* (1927–40) was devoted to 'Dream City and Dream House, Dreams of the Future, Anthropological Nihilism, Jung' (pages 388–404), while Convolute L contained 'Dream House, Museum, Spa' (pages 405–415). Often, Benjamin uses dreams to evoke a strange mix of technology, obsolescence and desire (or wish-fulfilment). This he could see in the very fabric of cities: in their museums, train stations, dioramas, factories, parks and squares, kiosks and casinos and, of course, arcades. His concern with these sites was to understand the dreams that had gone into building them.

For Benjamin, the dream is most vivid at the point of waking. Empirically, this means that he was most interested in those parts of the city that were being torn down or being altered, since it was as if people were waking up from the dreams that these spaces once embodied. Such places included, famously, the once fashionable arcades of Paris, but also the temporary structures put up for the great exhibitions of London and Paris. Indeed, Benjamin believed that the World Exhibition of 1867 installed Paris as the 'capital of luxury and fashion', where 'the phantasmagoria of capitalist culture attained its most radiant unfurling' (1935, page 166).[35] However, Benjamin also found modern dreams in museums (which contained artefacts, dreams, from the past) and railways stations (where there were dreams of travel). He uncovered dreams of previous generations in the ruins of the city: in their castles and churches. Like an archaeologist, he dug deeper and deeper into the historical and mythical layers of the city, to find the persistence of its dreams. Thus,

> **In order to understand the arcades from the ground up, we sink them into the deepest stratum of the dream; we speak of them as though they had struck us. (1927–40 [F°,34], page 841)**

Benjamin's exploration of the city proceeds dream-like, from particular moments or elements of the city: as he put it, he was 'on the track of things' (1927–40 [I1,3], page 212). In his dream-analysis of the arcades, Benjamin discovers many things about them. He finds that they were a site where both the prostitute and the *flâneur* found a home.[36] And the arcades are a place were the dreams embodied in commodities were stored, as elements in a dream. Each social figure or commodity could then be tracked in own right.[37]

Significantly, as with a dream, the arcades confused interiority and exteriority. In dreams, there appears to be no exterior world, while at the same time they are intimately connected to the generative processes of the body. Similarly, the arcades can be seen as an extension of the bourgeois interior into the public space of the street. The effect and affect of blurring interior and exterior worlds is phantasmagoric.

For Benjamin, the people, houses and streets of Paris themselves take on a dream-like quality. This quality is given further stimulus in those parts of Paris where labyrinthine street layouts persist. As in a dream, these streets can be entered and exited at many points, still and busy, with the constant possibility of finding oneself somewhere strange yet familiar; familiar, yet strange. An image that resonates strongly with fantastic stories such as G. K. Chesterton's 'An Angry Street' (1908) and Neil Gaiman's 'A Tale of Two Cities' (1993b).

Through the labyrinths of the city's streets, through the journeys undertaken, Benjamin would piece together the unconscious strivings of social and urban imagination. The analysis builds a labyrinthine picture of the thoughts that underlie city life: this labyrinthine quality exists in both space and time. Walter Benjamin cites Ferdinand Lion as saying:

> The most heterogeneous temporal elements thus coexist in the city [...] Whoever sets foot in a city feels caught up as in a web of dreams, where the most remote part is linked to the events of today. One house allies with another, no matter what period they come from, and a street is born. (1927–40 [M9.4], page 435)

Reading Benjamin's work is almost like walking through the city's web of dreams: at any point, you find places built out of different motivations or contradictory imperatives, sometimes side by side, sometimes in the same place (say, as one use blends into another). Indeed, the way that the city is produced as a web of dreams can properly be called city-work: work analogous to dream-work is carried out to produce city spaces in particular forms. In this sense, cities and city spaces are overdetermined, points of capture for chains of meaning, while at the same time they enable the shifting of affect from one part of the city to another, or to focus affect on one particular site. City-work constructs the city as a situation – or series of situations – in which desire can be dramatised. However, these desires are dramatised in disguised ways. Thus, commodities represent a desire, but not directly so. So it is, also, with physical infrastructure: homes, skyscrapers, fly-overs, subways, piazzas, and the like.

Figure 1.16 Trafalgar Square in July 2003.

It is possible to use London's Trafalgar Square as an example. The redesigned Square was opened to the public in July 2003 (Figure 1.16).[38] The open space now joins the National Gallery with Nelson's Column to provide a focal point for both London and the Nation.

the dreaming city   **51**

At three of the Square's corners, and from atop the column, imperial war heroes look down upon the tourists. From the Square, it is possible to see the Houses of Parliament and Admiralty Arch. In other words, one chain of meanings leads out to Britain's imperial and military past, another towards Parliament and the State. Associated with Parliament and the State is, for some at least, the notion of democracy. If we allow this idea to open up other chains of meaning, we can easily start thinking about the different kinds of politics that converge upon Trafalgar Square. Indeed, Trafalgar Square area has also long been associated with popular protest (Figure 1.17).

Figure 1.17
Protests at Trafalgar Square on 1 May 2003.

In London, on 1 May 1517, for example, there had been attacks against buildings associated with rich foreign merchants and craftsmen. London, before and after, witnessed other riots – such as, to name but a few, those in 1660 (by Charles II loyalists), 1736 (against Irish workers living in the area where Brick Lane is today), 1780 (by 'conservative' elements led by Lord Gordon), 1866 (for political reform), 1886 (by political radicals) and 1990 (against the Poll Tax) – though for markedly different (conservative/radical) reasons. (It has to be remembered that the so-called Left do not have a monopoly on either rioting or resistance.) Recently, on 1 May every year, Trafalgar Square has been used for demonstrations against capitalism and capitalist globalisation.[39] And it is also where thousands gather to celebrate New Year, much as they do in Times Square.

Trafalgar Square is undoubtedly overdetermined. To find anti-war protests in the same place as statues to imperial war heroes ought to be a puzzle. How is it that two contradictory affects and contents – peace and war, Imperialism and Radicalism – can occupy the same place? Perhaps because we are used to these paradoxes of space-work, the yes/no reality of the city – as an assembly of contradictions – loses its impact and people can no longer see the dreams that are embodied in their bricks and concrete, in their monuments and open spaces. As in dreams, fragments of the city are bound up in dense networks of affect, meaning and power, like flies caught in a web.

In cities, the most intricate structures of meaning and power are created, all of which are the points of articulation of many associative paths of meaning, all of which displace elsewhere the intensity of feelings that wish them into existence. For sure, an understanding of the city must trace the social relations that produce 'things' (from buildings to emotions) – as political economists since (before) Marx have pointed out. But now we must be sure that we understand that the 'things' (including commodities) that *make up* city life are also produced as phantasmagorias, spectacles of dream-like images in procession, with their hidden-in-plain-sight discourses of desire, desires that are thereby displaced along disparate paths.

Dreams, in Benjamin's view, are important precisely because they contain desires, desires that he believed the world needs to wake up to.[40] Indeed, this idea is central to Benjamin's revolutionary political philosophy. In the next section, we will see that Benjamin paid attention to dreams because he thought that they could be used to liberate their (hidden) desires. As importantly, dreams are an analytic model both for modern life and for the ways in which that city life is lived as if in a dream.

1.7 Cities are Lived as Dreams: Walter Benjamin on modernity and awakening

In this section, I will show how Walter Benjamin's understanding of dreams informed his theory of revolution as well as his analysis of modern city life. These are provocatively combined in one of his best-known works, 'One-Way Street' (written between August 1925 and September 1926).[41] 'One-Way Street' is presented as a series of fragments taken from signs that Benjamin has seen in the city. Each sign is used as a starting point for a train of thought – such as an anecdote or a metaphor, a dream or a short essay in political economy, pithy observations about this and that – as if Benjamin were one of Freud's patients free associating about a dream they had just had. This is the point. The sequencing and directions in 'One-Way Street' build, piece by piece, into a dream-like description of the phantasmagorias of modern city life, of the phantasmagoria of modernity itself. Its overall effect is to produce both an analysis and a radical critique of modern city life.

The fragments in 'One-Way Street' are bizarre, absurd, juxtaposed in odd, puzzling ways, their meaning not immediately apparent – and, when it is apparent, it becomes curious for being so obvious. The effect is deliberate. By juxtaposing these fragments in this way, Benjamin is attempting to bring seemingly unrelated things into a dialectical relationship. Through this process of dialectical imaging, Benjamin is seeking to use the tension between fragments to break them out of their (spatial) isolation, their (temporal) stasis. Through the shock of realisation, objects would be seen anew, as if the dust had been shaken off them. Awakening from the dream of modernity would, he hoped, allow the repressed wishes of history to be expressed and fulfilled. Even so, for Benjamin, 'dreaming has a share in history' and dreams have become colourless in modern times (see also section 1.2 above):[42]

> The dream has gone grey. The grey coating of dust on things is its best part. Dreams are now a shortcut to banality. Technology consigns the outer image of things to a long farewell, like banknotes that are bound to lose their value. It is then that the

> hand retrieves this outer cast in dreams and, even as they are slipping away, makes contact with familiar contours. It catches hold of objects at their most threadbare and timeworn point. (1927, page 3)

Despite this continuing catastrophe of history and modernity, dreams may yet yield something revolutionary and colourful about modern city life. But redeeming dreams will require new streets of meaning and value to be driven through them. 'One-Way Street' is directed towards this end.

The first sign in 'One-Way Street' is 'Filling Station'. Under this sign, Benjamin effectively introduces the work. He begins by talking about how the present is constructed out of facts, but facts that sterilise literary activity. The task of the critic, then, is to revivify this state of letters. In part, this is to be achieved through the use of opinions, which give writing both influence and the capacity to act. Such writing acts not in universal ways but through its specific, careful, accurate and effective application – much as one applies oil to a complex machine (hence the title of the fragment). It can be easily surmised, then, that 'One-Way Street' is the drop-by-drop application of criticism to 'the vast apparatus of social existence' (page 45). Through this process, the work can become a significant literary work. Another lesson to draw is this: these fragments are not assembled without connecting threads, without a perspective. These fragments are a rebus – locked in the puzzle are the desires and fears of the modern city. Let us proceed down the street.

The next sign is 'Breakfast Room'. Benjamin begins:

> A popular tradition warns against recounting dreams on an empty stomach. In this state, though awake, one remains under the sway of the dream. For washing brings only the surface of the body and the visible motor functions into the light, while in the deeper strata, even during the morning ablution, the grey penumbra of dream persists and, indeed, in the solitude of the first waking hour. (pages 45–46)

The grey penumbra of dreams persists for the dreamer who recounts dreams without having woken up properly. This is an analogy for modernity, whose dreams Benjamin described as grey and colourless. Benjamin is suggesting that modernity is recounting dreams on an empty stomach and remains, therefore, under the sway of dreams. This is significant for understanding Benjamin's attitude to modernity, dreams and revolutionary practice. He continues:

> The narration of dreams brings calamity, because a person still half in league with the dream world betrays it in his words and must incur its revenge. Expressed in more modern terms: he betrays himself. (page 46)

In a nutshell, it can be said that Benjamin is describing the alienation experienced by people in modernity. They betray themselves by articulating their dreams and the revenge that is wreaked upon them is that they have to exist in the dream-world of modernity. And from which they cannot awaken:

> He has outgrown the protection of dreaming naïveté, and in laying clumsy hands on his dream visions he surrenders himself [...] The fasting man tells his dream as if he were talking in his sleep. (page 46)

The moderns – after betraying their (innermost) dreams – are doomed to walk in a colourless alienated dream-world of modernity as if in their sleep. The problem is that they have no way of knowing that they are still half in league with the world of dreams. It is, therefore, the (revolutionary) task of the critic to shock the dreamers awake: to act as an alarm clock, to make the hammer strike the bell. For Benjamin, the desires of the sleepwalker in the modern city have to be materialised, but this is not as easy as it might be. The modern individual is perfectly capable of articulating a whole series of needs and wants, fears and anxieties. Indeed, the endless phantasmagoria of commodities taps directly into the *conscious* wishes of modern individuals. Unfortunately, although commodities seemingly embody and fulfil people's wishes, they remain alienated from people's actual, historical desires. In this sense, commodities become fetishes because they are worshipped, yet no-one knows why, or even what they stand for. It is as if the moderns are talking in their sleep: talking, asking, wishing, but only dimly aware of the meaning of the words they use.

> The nineteenth century is a space-time <Zeitraum> (a dreamtime <Zeit-traum>) in which the individual consciousness more and more secures itself in reflecting, while the collective consciousness sinks into ever deeper sleep. But just as the sleeper – in this respect like the madman – sets out on the microcosmic journey through his own body, and the noises and feelings of his insides [...] in the extravagantly heightened inner awareness of the sleeper, illusion and dream imagery which translates and accounts for them, so likewise for the dreaming collective, which, through the arcades, communes with its own insides. We must follow in its wake so as to expound the nineteenth century – in fashion and advertising, in buildings and politics – as the outcome of dream visions. (1927–40 [K1,4], page 389)

Though they can speak their wishes, the moderns have no way to make them real. The modern world becomes a never-ending cycle of dream-like figures – a phantasmagoria – none of which ever fulfils its promise. Fashions come and go: ever more rapidly, in ever more absurd forms. Buildings are put up and torn down, its façades become make-up in a clown's parade of architectural forms. This slumbering modernity presents a real analytical problem: how to waken the sleepwalkers and, as important, what are they to wake up to?

One might think that the poverty and injustice of the late nineteenth-century city was terrible enough to wake anyone from their slumbers. But Simmel's analysis of the modern city and its effects on human psychology (1903) suggests exactly the opposite to be true (see Introduction). From Simmel, Benjamin learns that the urbanite becomes indifferent to the shocks of city life and blasé about the sheer number of – absurd and surprising, dreadful and exciting – things that cities bring into close proximity. Despite the clarion calls against injustice and inequality, then, modern individuals are indifferent while at the same time becoming more and more obsessed with the rationality of their actions. As a result, city dwellers become subject to the revenge of dreams, for once they learn indifference and rationality, their desires and fears become a secret discourse in which everything conceals something else.

In Benjamin's analysis, 'dreaming' has two apparently contradictory meanings (perhaps following Freud's suggestion that dreams cannot choose between contradictory options). Dreaming describes both a state of sleeping and also a state of wakefulness. Both asleep and awake, consciously and unconsciously, the mind works *as if dreaming*. So, Benjamin examines dreams for a resource of (revolutionary) hope. He finds it in the idea that dreams anticipate awakening: dreams are a waiting room, where the sleeper lingers before escaping the clutches of sleep (1927–40 [K1a,2], page 390).[43] As with Freud, Benjamin did not wish the dreamer, moderns, to forget dreams on waking, but instead to wake up to the desires contained within them. This was Benjamin's hope: to use dreams to liberate as yet unexpressed, unfulfilled desires.

Analytically, then, Benjamin was concerned to discover and interpret dreams, both past and present. He sought these dreams both in artefacts – especially old-fashioned, obsolete and discarded objects, that Benjamin saw as embodying the dreams of their age – and in the places that housed, or contained, dream-images. By exploring how these urban dream fragments crystallised the wishes of earlier generations, Benjamin hoped to awaken modern city dwellers to their own desires. Though fully aware that desires and dreams could lead to violence and repression, Benjamin was optimistic. If he could bring the fragments into tension – through 'dialectical imaging', by putting the pieces side by side – Benjamin thought it would be possible to induce the right kind of shock that would wake up the moderns:

> The dialectical image is an image that emerges suddenly, in a flash. What has been is to be held fast – as an image flashing up in the now of its recognizability. (1927–40 [N9,7], page 473)

In 'One-Way Street', this revolutionary task manifests itself in the juxtaposition of ideas within observations, but also in the juxtaposition of observations. The 'Breakfast Room', for example, becomes a space which contains two apparently unconnected ideas: the pre-modern folk-tale and an interior space in the bourgeois home set aside for the timed and localised activity of breakfasting, now regimented by capitalist labour relations.[44] In the *Arcades Project*, alternatively, Benjamin suggests that the wake-up shock of recognition – or 'involuntary memory'[45] – might be induced through techniques of remembrance, partly by taking possession of the worlds of childhood (1927–40 [K1,1], page 388). Thus, the latent content of modern dreaming is allowed to break through the dream of modernity, like a terrorist's bomb shattering a bogus consensus.

Through dialectical methods, whether by use of images or memories, Benjamin was attempting to wake the modern world up, so that it could recognise and act on its dreams – rather than simply live in them, their eyes wide shut.[46] Even so, the idea of a phantasmagoria of dreams and the identification of the dream-work that goes into making dream-worlds offers much more than a way of saying that modern city dwellers are blissfully unaware of the true conditions of city life. It is on this point that we can conclude this chapter.

1.8 Conclusion: phantasmagorias, dream-work and the fulfilment of wishes

Both dreams and cities have to be made. In the production of both dreams and cities, it is evident that some real thinking has to take place. The landscapes of cities and dreams need to be interpreted, as their surface appearances – while revealing in themselves, while part of the story itself – are not the whole story. As for Robinson, in Patrick Keiller's film *London*, other meanings can be discerned in the landscapes of cities and dreams. There is no one dream that articulates the city, no one phantasmagoria that dominates or characterises a particular city's life, no one aspect of the city that defines its dreamings, its phantasmagorias. Instead, as with dreams, the interpretation of cities must trace the chains of association that emanate from urban fragments and combine this with an understanding both of their production in multiple social relations and of how the dream-city condenses and displaces meanings, emotions and power in its very form.

As I showed in section 1.2, there is – within both modernity and cities – a phantasmagoria of dreams. The phantasmagoria of modernity is dream-like: a procession of images, seemingly ever new, yet as Benjamin observed curiously the same. Dreams themselves are part of the procession of dream-like images through cities, part of the spectacle of city life. Dreams, thereby, lend to cities a sense of the phantasmagoric. Cities themselves contain a whole host of phantasmagorias and a vast array of infrastructures for manufacturing and maintaining them. Metropolitans are ever watchful for new images and ideas, new experiences and spectacles. Indeed, city folk enjoy their phantasmagorias, so much so that, according to Benjamin, they prefer to sleepwalk their way through the city and through history. In this way, the city is experienced as if it were a scene in a dream. The phantasmagorias of city life are, indeed, like scenes in a dream: not as passive backdrops, but as active constituents of the city itself. Each phantasmagoria drawing upon the raw materials of the city to create a dream-like narrative, composed out of specific, diverse elements: elements, such as commodities, or fashion, or dreams.

Following Freud and Benjamin, we can argue that dreams and cities are the guardians of the moderns' sleep: an elaborate play of remembering and forgetting; showing and disguising. In this understanding, both displacement and condensation work (to make dreams/to make cities) by using associative paths not only to combine and recombine thoughts, but also to off-centre both meaning and affect. Consequently, in the analysis of dreams and cities, one cannot necessarily expect to find causality in those elements that appear to be most meaningful, most powerful, most affecting. It is not that these elements should be ignored, but that they are only ever a starting point for analysis. Meanwhile, those apparently meaningless and overlooked elements of dreams and cities can also provide fruitful ways to access the networks of affect, meaning and power of both dreams and cities. In this understanding, dream-work, city-work and space-work describe different aspects of 'what gets done' to produce dreams, cities and spaces – and there are productive parallels to be drawn between them. One such parallel concerns the expression and concealment of the wishes that lie behind the production of dreams, cities and spaces. Significantly, the work that goes into making dreams, cities and spaces disguises not only their implicit networks of affect, meaning and power, but also the open secret of the motivations that lie behind them.

Benjamin wishes to put modernity in touch with its over-familiar secrets. Boldly stated, in his understanding, fragments of the city are *just like* fragments of a dream and, as such, can be interpreted in much the same way as Freud interpreted fragments in a dream. For Freud, the purpose of dream-analysis was to permit patients to 'wake up' to their hidden desires and fears – to facilitate, in the patients, a kind of transformative recognition of their own circumstances. Benjamin's patient, as it were, was modernity itself. Benjamin is not asking modernity to wake up *from* its dream of itself, but instead to get in touch with the wishes that the dream contains (as the epigrams to this chapter indicate). In other words, Benjamin's dream analysis of the modern city assumed that authentic wishes were woven into the making of the dream-world of modernity: the fulfilment of wishes was not to be found in waking up from the dream-world to reality, but in seeing the reality in the dream-world.

For Benjamin, the city becomes a dream-like site of memories and wishes. Indeed, he wished to evoke the quality of those memories and desires, though they remained as tantalising as a 'half-forgotten dream' (1932a, page 316). For my purposes, dream-work – and similarly space-work and city-work – are important because they speak to the ways one might track *both* fantasies through the objects of city life *and* the (dis)placements of affect in the city. Also, and reminding ourselves about 'issues of representability' and 'secondary revision', it is possible to discern the ways that the city is worked over by familiar stories – creating a coherence and intelligibility that might nonetheless betray the kinds of work that go into making the phantasmagorias of city life.

Neither Benjamin nor Freud believed that this *work* in affect and meaning was either simply free from, or simply released into, the world. In fact, a whole range of agencies line up to do exactly the opposite: to suppress affect, to censor thoughts, to disguise motivations. The agencies of the mind cannot be directly correlated with those of the social: it isn't accurate enough to say, for example, that 'the law' operates in exactly the same way as 'the super-ego' (or vice versa). Nevertheless, the rhetoric of dreams draws attention to social injustices and inequalities, if only by highlighting how far short reality is from delivering on its promised dream-worlds.

The promise and failure of dreams speaks to an injustice and inequality that is simultaneously personal and social. Both Freud and Benjamin saw in dreams the possibility that something vital could be liberated that would allow such injustices and inequalities to be both expressed and addressed. This possibility remained, like a half-forgotten dream, tantalisingly out of reach. It still does. But the legacy of these ideas can lead elsewhere – to an interpretation of elements in cities that combines, at each point, the personal and the social. In this way, it is possible to explore the ways the raw materials of the city form into wishes and to discern the patterns in these wishes, as they become concretised in cities and city life.

We are already in the midst of this analysis – dreams themselves, as we saw in sections 1.2 and 1.3, are part of the raw materials from which the dream-worlds of modernity and cities are made. There has been a clear indication, throughout, that these raw materials – and dream-thoughts – form into wishes. The next logical step in the argument, therefore, is to look in greater depth at how wishes form and gain expression in cities and in city life. But what to track? The urban literature casts up several possibilities: from prostitutes (on every street corner?) through to gamblers (congregated around slot machines in Las Vegas?).[47] Undoubtedly the desire for sex and wealth are part and parcel of the phantasmagorias of city life. Instead, however, I have chosen magic. The appearance of magic in cities is significant for a variety of reasons, which I will spell out in the introduction to the next chapter. What is important at this point is the link between dreams and magic and this has to do with the fulfilment of wishes.

What dreams tell us is that the popularity of magic should not be dismissed, as it might betray something of the hidden desires and fears that underlie city life. More than this, magic is a discourse and practice for articulating wishes and for making them come true. If dreams are the disguised fulfilment of repressed wishes, then the phantasmagoria of magic will reveal how cities and wishes are interwoven, through the magical expression of secret wishes. Yet the fulfilment of these wishes remains troubling, somehow unfulfilling. Magic consistently promises to fulfil wishes, yet there is also a caveat: *be careful what you wish for*. The wish, as if it were in a dream, can be uncannily, and ironically, thwarted. This idea has consequences for thinking about city life.

While dreams rely on networks of affect and meaning, on dream-work, the interpretation of magic tracks the occult, through occluded relationships. Occult beliefs and practices commonly suggest an alternative reality to the immediately experienced world, one often tantalisingly just out of reach. Real Cities, on the other hand, appear to be incontrovertible proof that these alternative realities do not exist. Magic and cities seem mutually exclusive, with magic and magical practices belonging to anywhere other than modernity and modern cities. By tracking magic and magical practices, we will see that this is very far from the case. We will discover that there is a phantasmagoria of magic in Real Cities, just as there is a phantasmagoria of dreams: an endless procession of magical images and practices at the heart of modern city life.

2
The Magic City
in which we are careful about what we wish for

2.1 Introduction: the magic of city life

At the core of dreams and magic are (unfulfilled) wishes. In the phantasmagoria of dreams, an endless procession of 'things' – from commodities to political rhetoric – offer to fulfil people's secret wishes – be they consumers or political audiences. So it is with magic. In this chapter, therefore, I am interested in exploring how it is that the phantasmagoria of magic appears in the city. Appreciating the magical practices that go into making city life will bring new issues to light. In particular, the emphasis on magic will involve a discussion of what is occult and supernatural about city life. Magic also implies a superstitious or primitive world-view. Often, this can seem quite at odds with both religion and science. Indeed, it has been assumed that modern city life would lead to the extinction of magical beliefs and practices. Examining this proposition lies at the heart of this chapter. I will show that magic itself consistently emerges from modern city life, as people seek ways to improve their conditions and influence their futures.

Magic, significantly, appears to lie between worlds, much as dreams do (which are between the waking and the sleeping, the personal and the collective, the conscious and the unconscious, worlds). In these in-between places, magic is overdetermined: occult and occluded relations produce magical practices and wishes. Tracking magical practices through cities allows us to see how occult relationships are, and become, significant for city life: partly as magic *circulates* through cities; partly through the ways that magic condenses and displaces 'things' in the city; and partly through the networks of affect, meaning and power that radiate out from magical practices.

Like dreams, magic forms a phantasmagoria that is consistently between worlds. In keeping with his understanding of phantasmagoria as a kind of threshold state that exists somewhere between its production and the ghostly visible world, Walter Benjamin also recognised the way that magic exists in-between worlds:

> Threshold magic. At the entrance to the skating rink, to the pub, to the tennis court, to resort locations: *penates*.[1] The hen that lays the golden praline-eggs, the

machine that stamps our names on nameplates, slot machines, fortune-telling devices, and above all weighing machines (the Delphic 'know thyself' of our day) – these guard the threshold. Oddly, such machines don't flourish in the city, but rather are a component of excursion sites, of beer gardens in the suburbs [...] Of course, this same magic prevails more covertly in the interior of the bourgeois dwelling. Chairs beside an entrance, photographs flanking a doorway, are fallen household deities, and the violence they must appease grips our hearts even today at each ringing of the doorbell. Try, though, to withstand the violence. Alone in an apartment, try not to bend to the insistent ringing. You will find it as difficult as exorcism. Like all magic substance, this too is once again reduced at some point to sex – in pornography. Around 1830, Paris amused itself with obscene lithos that featured sliding doors and windows. (Benjamin, 1927–40 [I1a,4], page 214)

For Benjamin, there is a magic to the threshold, as an intermediate and indeterminate space where different worlds meet – as between interior and exterior worlds. In Benjamin's analysis, fortune-telling machines cross between a person's present and their future; bourgeois interiors between the world outside and the world of the family. This meeting of worlds has different consequences: some thresholds, for example, are guarded, while others are marked by rituals that allow entrance or exit to be gained. What is magical, though, is the possibility that something else will *emerge* through the encounter between what's on one side of the threshold and the other.[2] Benjamin's example is, in a somewhat Freudian flourish, about sexual motives and prohibitions. The pornographic use of thresholds, in sliding doors and windows, reveals bourgeois sexual desires for what they are. What is magical, at this threshold, is the possibility that sexual prohibitions might be exorcised. From this example, we can see that threshold magic does 'double duty' for Benjamin, both as a site of critical analysis that reveals a secret (wish, desire) and, through the transformative recognition of these secrets, these sites also have the potential to be redemptive (a bit like dreams: see section 1.7).

The emergence of magic in thresholds doesn't just involve the meeting of different worlds, it suggests that magic itself is in contact with other, occult worlds. The phantasmagoria of magic brings with it a sense of other-worldliness, of the hidden networks that constitute modernity and the city: that is, of the occult relations of modern city life. These relations are powerful even as they are intangible, for magic relies on the idea that occult relations have force in the world – even while they are intangible, invisible, mysterious and so on. It is perhaps for this reason that Benjamin perceives the other-worlds that cling to ordinary household objects such as chairs and family photographs, doorways and fortune-telling machines: all, for him, seem to call forth fallen deities or family revenants from another time and place – gods and ghosts that are almost impossible to exorcise. Yet, magic also offers the prospect of utilising these occult forces, of touching the intangible, of seeing the invisible worlds beyond perception. And magic is infused with desire: to 'open sesame', to see what cannot (yet) be seen, to make things happen. Yet, magic – like the dream – is not an innocent world. In the worlds that magic calls up, there is as much potential for good as evil. As they say, *be careful what you wish for*.

In **section 2.2**, I will explore the supposed distance between the urban mind and the superstitious mind by looking at a less well-known article by Robert Park, a leading member of the Chicago School of Sociology. In this piece, Park attempts to prove that belief in magic is alien to cities and also that the mentalities that city dwellers necessarily develop are antipathetic to beliefs in irrational superstition. Magic, he asserts, cannot survive in cities. Yet, there are those who would argue that cities are just as magical, spiritual and other-worldly as anywhere else. In **section 2.3**, we will look at the case of New York to find out whether magic withers in the modern city, as Park insists. We will see that, for some, cities are as good a place for magic as anywhere else. Indeed, it can be argued that magic necessarily flourishes in cities. The example of New York shows, moreover, that magical powers and practices do not exist somehow independently of city life. This weaving of magic through, and also the invention of magics within, cities and city life is explored in **section 2.4**. Here, New Orleans and Voodoo is exemplary. I show that magic adapts and mutates, using urban social relations as well as the fabric and commodities of cities to create new ways of mediating (occluded, occult) powers.

For Park, the dominance of cities by abstract and calculative business practices places them outside magical influence. There are those who would suggest that urban economies can learn much from magic, including Voodoo. These arguments are explored in **section 2.5**. The point here is not only that magic has been used to maintain prosperity and wealth, but also that wealth and prosperity is itself somehow magical. Arguments such as this resonate with certain beliefs and practices that seek to manipulate space to positive effect. Therefore, the influence of *feng shui* on how space is produced is examined in **section 2.6**. Magic, however, is not necessarily benign. In the final substantive section (**section 2.7**), I examine the case of 'Adam', an as yet unidentified boy who was the victim of a ritual murder. Adam's case throws up two more aspects of magic and city life: on the one hand, it highlights a global trade in magic; on the other hand, it sickeningly demonstrates the dire consequences of this trade for the most vulnerable. In **section 2.8**, by way of conclusion, I talk about the wishfulness, the dangers and the disappointments, and the occult relations that constitute modern city life.

2.2 Civilisation and its Discontents: the science of the city versus the powers of magic

At first glance, it seems that magic is commonly pushed to the margins of the city. For example, Benjamin can only find magical devices – such as one-armed bandits and fortune-telling machines – in suburbs, in tourist sites and beer gardens. He was not alone in assuming that cities force magic to their fringes. For some, cities are civilisation embodied – the very antithesis of primitive beliefs and practices. If this is the argument, then it would be nice to have some measure of civilisation that would incontrovertibly demonstrate that cities are indeed the highest stage of human development. For Robert Park, this measure of civilisation would be found in the *distance* between civilised mentalities and primitive mentalities.[3] Modern mentalities, in his view, are less (and less) susceptible to superstitious beliefs and beliefs in magic. And, so, the belief in magic would be the litmus test of civilisation:

> magic may be regarded [...] as an index, in a rough way, not merely of the mentality, but of the general cultural level of races, peoples, and classes. (1925b, page 131)

Park's arguments bear a remarkable similarity to those of Sigmund Freud and this is worth commenting upon; the differences (and similarities) between their arguments lay the foundations for questioning the idea that urban mentalities are characteristically rational, calculative and scientific (an argument that can be triangulated with the work of Simmel). For Freud, civilisation is about repression: its victory is over human instinctual behaviour. For Park, this is essentially about intellectual progress and enlightenment. In both views, there are racist assumptions, but the *pas-de-deux* of science and magic in their arguments reveals that civilisation – and the modern city – is very far from free (or freeing itself) of magic. Indeed, alternatively, we might wonder whether cities are being produced by magical powers, influences and practices. So this section opens up the possibility of seeing the magic of modern city life.

Park was keen to show that science and city culture represented higher stages of cultural development than magic and folk culture. To begin with, he tried to draw a clear distinction between science and magic. However, each criteria he applied tended to blur the boundaries between them: if science was about control over the external world, then so was magic; if magic assumed certain connections between things, then so did science; if science was about action interrupted by reflection, then so was magic; if magic was about the formulation and expression of a wish, then so was science.

Unperturbed by his failure to establish an absolute difference between science and magic, Park decided to apply his measure of civilisation to a 'living laboratory', a place where the dichotomy between the city culture and folk culture could be validated. If cities such as Chicago and New York were marked by their openness to the world, Park argued his laboratory would have to be a place were a culture had developed uncontaminated by the world around it (as if Park were Darwin on the Galapagos Islands). To test his hypothesis, he selected the belief in obeah on English Caribbean islands. On these islands, Park claimed, people lived in relatively isolated, untouched social units. By examining their cultures, he argued, the inherent, primitive predispositions of peoples/races could be revealed. This unquestionably flawed, racist logic propels Park forward. The belief in obeah, he insisted, demonstrated that West Indian island cultures were distinct from – and *less civilised* than – urban American cultures.

> **The uncivilised man [sic] enters, so to speak, into the world about him and interprets plants, animals, the changing season, and the weather in terms of his own impulses and conscious purposes. It is not that he is lacking in observation, but he has no mental patterns in which to think and describe the shifts and changes to the external world, except those offered by the mutations of his own inner life. (1925b, pages 125–126)**

In this view, there is no mediation between the individual's inner life and the external world. Obeah – and magical practices in general – therefore work by allowing the inner life of the individual to be *projected* on to the external world in such a way as to bring about desired-for outcomes. Indeed, the external world is interpreted only in relation to the inner purposes of the individual. On the other hand, this means the individual is open to direct influence by that external world.[4] It is not that these beliefs go untested against reality, but that failures of the world to live up to magical predictions are attributed to magical causes, such as the influence of the dead,[5] of things, or of others able to use magic (whether directly or through witches). For Park, such beliefs left people susceptible both to irrational fears and to a sense of the strangeness of things. The world around is not only determined by processes beyond explanation, but that other world is often perceived to be malign, savage. The world is full of demons, vampires and ghosts. In such a world, people expend a great deal of effort warding off evil and protecting themselves against the malevolent wishes of others.

Magic, further, is embedded in a whole series of practices that are emotional, expressive, gestural, dramatic, impulsive and performative. These practices are deployed by individuals in pursuit of the fulfilment of their wishes. Science, for Park, is different: not in that it does not wish, but that these wishes are formulated through reasoned debate in public and are enacted by democratic public process:

> Science formulates its wish, consecrates it – through the solemn referendum of a popular election, perhaps – and writes it on the statute book. (1925b, page 128)

We can see that Park uses politics (elections, legislatures, laws) to exemplify the scientific mentality. This is because the public domain demands a different attitude towards wishes. In this formula, mass societies articulate their wishes through informed debate and representatives of the masses serve the wishes of the majority. It is this that distinguishes urban America from island Caribbean. The effect of this collective wishing is to install a different mentality in the citizens of mass societies.[6] This mentality can be described as scientific. Thus, from this perspective, cities create a scientific mentality because they demand of their citizens a capacity for detachment, for critical reflection, for rationalism, for logical thought, for calculation and for an ability to understand the relationship between means and ends – all key features of science. Indeed, cities are a visible sign of the successful application of science and, therefore, an extension of an inner life that is rational, calculative, deterministic. You can see it in the concrete, glass and metal – and also in the very machines that are used in cities and used to build cities. In Park's scheme, magic can only survive *outside* the modern city.

Though there are many objections to this argument, at least one being its blatantly racist assumptions about (supposedly uncivilised and backward) Caribbean and African belief systems, I would like to explore two in more detail. The first becomes clear when you trace Park's argument back through Freud's analysis of magic in 'Totem and Taboo' (1913). Freud gives us an alternative sense of the relationship between the individual, wishes and collective beliefs. The first objection, on this basis, is that the modern city, and its citizens, can abandon completely their impulsiveness, irrationalities and emotions. The second objection concerns Park's assumption about the untouched nature of the Caribbean islands. By tracking magical beliefs and practices, we can see that magic has been important not only in the phantasmagorias of city life but also in how people have sought to influence and control cities, magically and otherwise.

There are interesting correspondences between Park's and Freud's arguments, so much so that it is possible to argue that Park borrowed heavily on Freud's argument that there is a diminution of taboo in modern civilisation.[7] Like Park, Freud sees magic as a way of controlling the world, of getting you what you wish for (including the ability to harm others), or for protecting oneself against the evil impulses of others. Also, both Park and Freud see magic both as a set of practices and also as a belief that certain objects (by association) can influence matters. Nevertheless, Freud is concerned with the emotional dynamics underpinning magical beliefs. Freud sees modern individuals as less expressive, less intensely emotional, than other peoples, since they are also *more repressed*: that is, that moderns are (too) good at renouncing certain desires (hence the common warning: be careful what you wish for).[8] However, for Freud, the civilisation of affect does not necessarily mean the extinction of a belief in magics. Indeed, the continuing need to deal with repressed or suppressed affect may lead to firm beliefs in magical influences and the magical properties of things.[9]

Freud can be read both against himself and against Park as providing a case for the persistence, rather than abandonment, of the belief in magical properties and practices. Reading Freud in this way adds two further insights to an understanding of the role of magic in modern city life: the first concerns the space-time of magic, while the second is about the emotional entanglements bound up in magic.

In Freud's understanding, magic works primarily through two kinds of spatiality: through similarity (the is-like-ness of things) and through proximity (the contamination of things). An obvious example of this might be the Voodoo doll (Figure 2.1).[10] Not only is it shaped to resemble a person, but magical influence is gained by using things associated with an individual — such as things that have come into contact with the person or parts of the body: for example, clothes and hair, respectively. The influence of magic also collapses space-time in specific ways, such that 'real' space-time is of no importance. The Voodoo doll, thus, has instant effects at a distance.[11] Magical influence also jumps spatial scale: parts are used to influence or control wholes. Thus, pins are stuck in the genitals of Voodoo dolls to guarantee the sexual domination of one person by another. Underlying this is a belief in occult relations between things: relations that can both collapse distances and times and also increase the scale of influence and control.

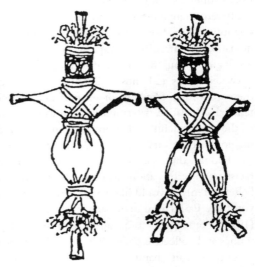

Figure 2.1
New Orleans stick dolls, designed by C. M. Gandolfo. Hair of the person to be affected is used to make the doll. The desired influence is written in red ink on parchment and placed inside the doll. White or black pins are employed in the spell.

For Freud, understanding magic is a lot like understanding dreams. Each element in magic is overdetermined: both acting as a nodal point within multiple networks of meaning and power, and also capable of producing a wished-for outcome. Each element is situated within the occult relations through which magical influences work. The work of magic, or **magic-work**, is designed to guide and focus wishes to have causative effects using knowledge of these occult relations (Freud, 1913, page 141).

However, if the success of magic-work relies on distilling and intensifying wishes, then problems arise where conscious and unconscious wishes are in conflict; or, worse, if there are a whole series of conflicting wishes within the mind, be they conscious and/or unconscious.

This leads to the second issue: the intense ambivalences bound up in magic.[12] Like dreams, the wishes seeking to be expressed in, and fulfilled by, magic can be disturbing, as they are often associated with forbidden acts, sins, of various kinds (following Freud, 1913, page 146). The stronger the prohibition on the wish, the more complex and demanding the magic required to fulfil the wish. But, if the wish is fulfilled, the more ambivalent will be the emotions surrounding it. Remember: what goes around comes around; be careful what you wish for. Thus, if you murder someone by magic, be worried that the ghost of your victim may return to haunt you forever. It is perhaps for this reason that much magic is devoted to defence: the use of rituals, prayers, charms and spells to protect against evil, evil spirits and the evil spells of others.

So, it is worth remembering that science and technology, as Park describes it, is part of the wish fulfilment – the magic-work – of the city. Problems arise where people in cities are intensely ambivalent about their wishes, and/or where there are intense conflicts over wishes and their fulfilment. Yet this is not the end of it: the rational production of city space cannot create a city that fulfils people's dreams precisely because, as Freud implies, their express wishes have ambiguous, ambivalent, contradictory, indeterminate, insatiable and multiple sources. As we follow magic through the city and witness how it produces city life in various ways, as we use magic to explore the emotional life of cities, we should be wary of reading it as pointing to a coherent, stable or singular wish. Instead, magic will embody the intense ambivalences surrounding wishes and sometimes give us a glimpse of the forbidden. By tracking magical beliefs and practices, moreover, it will be possible to trace out the occult or occluded connections between cities. An example of this is to be found in Park's own writings.

Park notes that West Indian magic is to be found in parts of American cities, for example on 135th Street in New York. For him, the appearance of magic in cities is *both* external to the city *and also* constantly being undermined by the civilising, scientific, influence of city life. If Park had been less anxious about distinguishing between the West Indies and urban America, he might not have overlooked the long and intimate history of influence and connection between the two. In the midst of a discussion of obeah rituals, Park himself observes that:

> Among the instruments of obeah in the possession of police in Trinidad recently were a stone image, evidently of Hindu origin, and a book of magic ritual published in Chicago, which pretended to be, and no doubt had been, translated originally from the writings of Albertus Magnus, the great medieval writer on magic. (1925b, page 133)

On Nevis in 1916,[13] though deaf and bedridden, Rose Eudelle was convicted of practising obeah. According to Park (pages 135–136), her witchcraft was practised mainly through correspondence and letters were found that connected her to clients on many islands and even one in New York City. Examples such as this, and others that Parks lists, demonstrate that the contacts between obeah men and women and their clientele stretched far and wide. Far from being untouched by the world, obeah – magic – throws light on an alternative globalisation: magical stones from India, letters sent back and forth to New York, books imported from Chicago, magical beliefs from Europe. Indeed, this has been a long-standing feature of the islands, and was not somehow simply 'new' in 1916. The islands' historical connectedness did not even begin with the earliest *European* colonisations of the American continent and the Caribbean, particularly by the Spanish and French.

Nonetheless, from the earliest moments of European colonial contact, various West African religious beliefs were mixed together, and mixed again with various Catholic practices, to make syncretic varieties of obeah or Vodou (we will return to this story below). Instead of seeing the islands as somehow frozen in time, backward and primitive, they are nodal points in the production of new (syncretic) magical practices and beliefs. What I am trying to establish, at this point, is that Chicago and New York were – and are – at the frontier of a globalisation of magic that was – and is – centred on the Caribbean (and not vice versa). It is ironic, therefore, that Park should consider the English Caribbean as a suitable laboratory for considering isolated cultural development: he might have been better off thinking about Chicago University's campus (see Nast, 2000). Nevertheless, if Park is right then New York's civilising influence would have been too much for the practitioners of magic on 135th Street. Progressively, the magic would have died out of the city. It is worth a visit to New York (Figure 2.2), then, to see if there's any magic left.

2.3 New York's Magic: supernatures, spirits and Vodou

For some observers, such as Park, cities are the last place you'd find anything magical: full of cold concrete, clear glass and heavy metal – all created and fashioned by rational humans and their clever sciences. For others, however, the city – in its very use of natural elements, in its

Figure 2.2 The magic of the city? Times Square at night.

Figure 2.3
Concrete, glass, metal: the magic of New York buildings.

wondrous physicality – creates a magical environment. Such an idea is far from popular or common, but a few within the 'new age' literature (as it is popularly known) have sought magic in the city. For example, the urban shaman Chris Penczak firmly believes the city is as magical as anywhere else:

> Many of those in rural settings think they have a monopoly on what is natural, magical, and spiritual. They don't. Everything is natural. Many in the city think they are confined to a nonspiritual setting, filled with unnatural and harmful forces. I thought so for a while. I felt disconnected from the world in the city, until I learned to forge new connections with the world around me. Again, everything is natural [...] Magick is hiding everywhere, waiting to be found. (2001, page xiv)[14]

Everything is (potentially, at least) magical – including both the city as a whole and the elements that go to make up cities, such as concrete, glass and metal (Figure 2.3). For Penczak, cities are a threshold space – connecting and combining fields of energies, elements of nature (i.e. earth, air, water, fire) and also spirits. Magic is everywhere, according to this logic, in all elements that comprise the city: in its machines, its buildings, its animals, its districts and neighbourhoods. In the first instance, the task is to become sensitive to the magical properties of city things:

> The city is a powerful landscape of magick, filled with secrets and energy for those who know where to look. (2001, page 3)

In *City Magick*,[15] Penczak unfolds a complex cosmology. He draws upon a wide range of belief systems to outline the various magical properties of cities: its elements (air, water, fire, earth); its totems (from spiders to pigeons, ants to crows); its spirits (everything has a *deva*: the mechanical spirits include planes, trains and automobiles); its Gods (such as electricity) and archetypes (from building types to medicine, from computer technology to music). In every element of the city, Penczak discerns a natural magick.[16] Indeed, by concentrating and combining these elements, cities are intense potential spaces for the creation of new magicks. There is a geography to this magick, for the energetics and spirits that underlie the visible world cross and swirl at nodal points:

> The planet is covered with different sets of energy lines and grids. They are like the nervous system and acupuncture meridians of Earth. Many of the lines cross at certain points in each grid, creating an energy zone, or vortex. Perhaps it is the other way around and the vortex creates the energy lines. (2001, page 4)

It does not matter that Penczak is uncertain as to whether the nodes create the network or vice versa. What is important is that these networks of energy and spirit have within them sites of intensity. This has two consequences. First, the focus and form of intensity within any one city provides it with its own special atmosphere. Thus, for Penczak, New Orleans and London have particularly magical atmospheres, alerting sensitive people to the mystical, invisible worlds that lie beyond the purely visible. Second, buildings play a significant part in this geography of intensity, partly because they represent nodal points in fields of energy and power. With both these points, London psychogeographers – such as Sinclair and Keiller – are liable to agree. As a critical discourse of power, many psychogeographers believe that the buildings of the powerful are built (deliberately) upon key nodes in the energetic system of the Earth: famously, this includes both the British Museum[17] and Canary Wharf (Figure 2.4).

Figure 2.4 At the intersection of lines of power, Canary Wharf.

Figure 2.5 The Empire State Building – the energy and spirit of New York.

Penczak is less sceptical about powerful buildings, for him they represent the energy and spirit of the city. Instead, Penczak concludes that there are many maps of reality. Penczak distils these into a cosmos comprised of three distinct worlds: of the sky, of the land, of the earth. Perhaps unsurprisingly for someone who lives in New York, the skyscraper is a significant building in his cosmology (Figure 2.5). For Penczak, the skyscraper vitally connects these three worlds: first, it reaches into the sky and the realm of the gods; second, into the earth and the underworld of the past, the dead, sewers and the city's services; third, the building itself is a multi-storied world, with flows of people and energy into and out from it.

The energy, spirit and physical form of the city offers the possibility both of new shamanic understandings of the cosmos, and also of attaining new (deeper) states of consciousness. With these greater states of perception, the urban shaman is able to take elements in the city and rearrange them to produce magical effects and influence. Thus, magic-work is often accomplished using urban elements such as concrete or glass or even graffiti: a good example of this is the sigil, a magic sign (Figure 2.6).

Penczak produces sigils by distilling the graffiti he sees in the city (Figure 2.7) or by joining key sites on a map of Manhattan (and so on). Using sigils, Penczak seeks to

Figure 2.6 A sigil using a pentagram and the Greek letter sigma, King's Cross, London.

Figure 2.7 Sigil material? A London 'eye'.

the magic city 69

influence and control the world: for example, he claims great success for a sigil, made by joining home, work and a parking place on a map, that ensures he always has a spot to park his car.

Instead of seeing the city, science and civilisation in opposition to or more advanced than magic, spirituality and the supernatural (as Park and others do), Penczak wishes city dwellers to (re)connect to the worlds beyond visible reality:

Reality is mutable. That is one of the basic facts of magick [...] Shamans and witches see the world for what it is – an illusion. (2001, page 75)

Seeing through the illusion means that shamans and witches can work to produce a Utopian city that has been (and always will be) out of reach of conventional (rational and scientific) ways of thinking about the city. For Penczak, by using shamanic rituals and practices that draw inspiration and knowledge from the city (including some developed from Voodoo: see below), a new harmony and settlement can be achieved with the unseen and secret forces creating life.

There is, as you'll have spotted, a new-age sentiment underlying Penczak's magical New York. The underlying suggestion is that everything is natural and, as a result, potentially in balance and harmony. In this view, contemporary city problems stem both from the inability of city dwellers to get in touch with the worlds beyond the visible and also (therefore) from the imbalances between things, including people, in the city. Once urbanites have got in touch with the supernatural realms, they will be able to use their knowledge to improve city life *as they wish*. This view of magick, common in new-age thought more broadly, corresponds directly to Freud's definition of magic (1913, page 135). In magical beliefs, Freud argues, there is a direct correlation between people's inner worlds (including their desire and faith) and their ability to use magic to exert control over the external world. For Freud, as there is no intermediation between internal and external worlds, magic is the lowest form of belief. Higher, he thought, was sorcery.

In sorcery, the other-worlds of spirits and magic were treated as if they were like the human world (even if these spirit worlds were believed to be more real or more powerful). As a consequence, in sorcery magical influence is gained through very human means: through bribery, appeasement, reparation, intimidation, disempowerment, subjugation, deception, misdirection, seduction, renunciation and/or begging of various kinds. In this view, the natural state of the cosmos is far from one of balance and harmony. Instead, the spirit world is just as conflictual, mischievous and self-interested as the human world and the outcome of contacting spirits is far from certain: indeed, be careful what you wish for. New York contains these cosmologies too – an example is vodou. In Park's view, there would be no place for Caribbean Vodou beliefs in New York. We can get a sense of whether Park is right or wrong about this from the story of a New York Vodou priestess, Mama Lola – as told by anthropologist Karen McCarthy Brown (2001).

In 1962, while in her early twenties, Mama Lola (also known as Alourdes) migrated from Port-au-Prince, Haiti.[18] She describes the hopefulness and disappointment of this to Brown:

I'm going to that city – lot'a star! Beautiful! Oh Boy! I don't think I'm going to need no spirit in New York. And I was wrong! (Mama Lola, cited by Brown, 2001, page 71)

Brown describes the life of poverty that awaited Alourdes in New York (2001, especially pages 124–133). As far as possible, moreover, Alourdes attempted to avoid the official world of public authority. Survival for Alourdes meant keeping out of sight. Protection would have to come, instead, from the spirit world. As a result, Alourdes' life was a constant struggle to make do on very little. Her sensitivity and training in Vodou would help, but it was not enough to put food on the table. She moved from low-paid job to low-paid job and there were always financial problems. Nonetheless, Mama Lola has built a reputation as a Vodou priestess and now has a devoted congregation. In its practices, Haitian Vodou is adapted to the difficulties of life in New York, both in its beliefs and its practices.

Indeed, Vodou beliefs are characterised by their adaptability and flexibility, as is Voodoo more broadly. Thus, it is important to realise that Haitian Vodou (and Vodoo) is closely linked to Catholicism.[19] Vodou/Voodoo is made up of a pantheon of spirits that closely resemble Catholic saints. As Brown puts it, while describing Mama Lola's Vodou altars and shrines:

> The lithographs included several images of the Virgin Mary and one each of Saint Patrick with snakes at his feet; Saint Gerard contemplating a skull; Saint James, the crusader on his rearing horse; and Saint Isidore, the pilgrim kneeling to pray by a freshly plowed field. These I recognized as images of the Vodou spirits. Each of these spirits has both a Catholic and an African name: Mary is Ezili, the Vodou love spirit; Saint Patrick is the serpent spirit, Danbala; Saint Gerard is Gede, master of the cemetery; Saint James is the warrior Ogou; and Isidore is the peasant farmer Azaka. (2001, page 3)

Vodou/Voodoo is capable of assimilating many beliefs, not just the pantheon of Catholic saints. This makes it very flexible and enables it to adapt to 'local circumstances' with great ease. Rituals and performances always have multiple meanings and can be read in a variety of ways by everyone concerned. There are prosaic and spiritual dimensions to each act in a Vodou ceremony. Indeed, as Brown argues, 'social drama is never far from the surface in the religious drama of Vodou' (2001, page 57). This multiplicity of possible meanings, as well as the uncertainty of outcome, is closely allied to the difficulties and perils of leading migrant lives in New York City. It is a belief system adapted to peril. Indeed, Vodou spirits themselves are often perilous: capricious, arrogant, spiteful – quite unlike the saintliness of Catholic figures. Brown puts it like this:

> It is no exaggeration to say that Haitians believe that living and suffering are inseparable. Vodou is the system they have devised to deal with the suffering that is life, a system whose purpose is to minimize pain, avoid disaster, cushion

loss, and strengthen survivors and survival instincts. The drama of Vodou therefore occurs not so much within the rituals themselves as in the junction between the rituals and the troubled lives of the devotees. People bring the burdens and pains of their lives to this religious system in the hope of being healed. (2001, page 10)

It is worth saying that such a belief system did not begin in New York, but lives led in poverty and pain have given Haitians good reason to believe that life and suffering are one and the same. For Alourdes, New York was hardly disabusing her of such an idea. Nonetheless, Vodou does do something else: it gives women significant status in the community (see also Marie Laveau's story in the next section). Through a system of 'donation' (although this is often expressly demanded by the spirits), it permits money and things to be given to the priestess, even if she herself never asks for payment of any kind. As Brown argues, Vodou provides women with 'a way of working realistically and creatively with the forces that define and confine them' (2001, page 221).

Vodou is a religion closely bound to the lives of those in New York, not just as a means of survival, but also in its practices. Thus, for example, ingredients taken from the city are used to make charms. These ingredients are closely connected (whether by similarity and condensation, contiguity and displacement, as Freud would say) to the problem for which the charm is a solution. In the city, new combinations of new elements are mixed up together to make the charm or potion. There is science involved here, using theories of the body and the soul derived from already existing Vodou beliefs or from other belief systems, such as Christianity. But there is also pragmatism and experimentation. It's flexible.

It turns out, contra Park, that New York – the modern city – provides fertile ground for magical beliefs and practices. From Penczak, we learn that New York can provide a model for new cosmologies as well as evidence of the other-worlds of energy, spirit and magic that make up reality. Meanwhile, Vodou provides a resource for Haitians in New York because it helps maintain social, cultural and religious ties with Haiti. From Mama Lola, we learn that Haitian migrants draw strength from Vodou because it can adapt to – and make use of – whatever resource New York has to offer. In both these stories, magic finds a place within New York – both as a belief system adapted to the form and life of the modern city and as a resource for magical intervention.

Magic, quite clearly, is to be found *within* New York. But I want to argue that magic is not just *in the city*, but also *of the city*. By broadening our appreciation of Vodou to Voodoo, it is possible to show both that magical beliefs – as a phantasmagoria – circulate widely through cities, and also that new magical beliefs are fostered and created within cities. Voodoo in New Orleans will exemplify both these points.[20] Two histories are important in the creation of Voodoo in New Orleans: first, there is the history of slavery in the Caribbean and the American South; second, there is the deliberate use of Catholicism and the spaces of the city by slaves to enable them to take control of their own lives. One history is about the crossing of beliefs across the Atlantic – the Voodoo Atlantic. The other is about the crossing of African and Catholic beliefs together to create a new, specifically urban, religion. It is to New Orleans that we now turn.

2.4 Syncretic Cities: New Orleans and the Voodoo Atlantic

In New Orleans, the particular mixing of West African and Catholic beliefs led to a set of magical practices based on herbal remedies, divination (foretelling the future) and casting spells. In these practices, rituals were used to formulate and express wishes, while various elements were utilised to create spells. These elements were – and are – highly eclectic, everything from snakes and skulls to Malibu rum and cigarettes, all designed to appeal to the specific Voodoo god (or saint) capable of delivering the desired outcome (Figure 2.8).

The tale of Voodoo in New Orleans is deeply connected with a particular site, Congo Square, and with Marie Laveau (the name of both a mother and a daughter). Through the stories of Congo Square and Marie Laveau, I want to convey both how certain ideas travel across the Atlantic and also how they are adapted and practised syncretically. Significantly, New Orleans today is renowned the world over for its Voodoo histories (Figures 2.9, 2.10 and 2.11).

The first place to visit in our tale of New Orleans' Voodoo is Congo Square or, as it has been called in the past, *Place des Nègres*.[21] Congo Square began life as a market-place in the years after the establishment of a French colony in Louisiana in 1699. Initially, the colony could barely sustain itself and relied heavily on trade and co-operation with local native American tribes and bands. Indeed, in this period, the garrison was abandoned at least twice as the French settlers were forced to take refuge with the local tribes, mainly

Figure 2.8 An informal Voodoo shrine in New Orleans.

the magic city

Figure 2.9
A Voodoo bar.

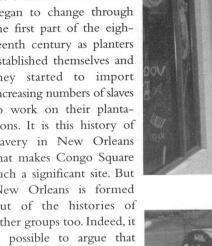

Figure 2.10
Marie Laveau's House of Voodoo.

the Natchez. The situation began to change through the first part of the eighteenth century as planters established themselves and they started to import increasing numbers of slaves to work on their plantations. It is this history of slavery in New Orleans that makes Congo Square such a significant site. But New Orleans is formed out of the histories of other groups too. Indeed, it is possible to argue that there were, and are, several New Orleanses: one, certainly, is American (that 'coarse mix' of different peoples); one French; one Catholic; but another is distinctly African.

New Orleans itself was founded in 1718 as the capital of colonial Louisiana (Hall, 1992, Chapter 5). Its survival, according to Gwendolyn Hall,

> was due not only to African labor but also to African technology. The introduction from Africa of rice seeds and of slaves who knew how to cultivate rice assured the only reliable food crop that could be grown in the swamplands in and around New Orleans. (page 121)

The variety of rice, *Oryza glaberrima*, and the people who knew how to grow it had come from West Africa. Tracing the origins of these peoples is obviously somewhat fraught, but correlating rice, technology and Voodoo suggests that these were predominantly Bambara, Dahomey, Yoruba, Ewe, Fon and Ibo peoples. Indeed, studies of Voodoo in contemporary Haiti and Benin have suggested strong parallels between these and Dahomey, Fon and Yoruba beliefs.[22] There is, in this view, a Voodoo Atlantic.

Figure 2.11
The 'Haunted History' tours leaflet.

As is well known, the Atlantic passage from West Africa to the Caribbean and the American South was brutal, genocidal. Africans were stripped of everything they owned: few possessions crossed the Atlantic, names were changed and the use of African languages strictly forbidden. It is testament to the strength, courage and ingenuity of African peoples that they were able to maintain their belief in 'spirits' and their cultural and religious practices, which in/famously include dancing to the beat of drums (see below). What is clear is that these traditions were taken across the Atlantic and integrated with other beliefs under slave conditions. Diverse cosmologies, rituals and forms of magic-work were combined into a flexible and open system of beliefs, namely Voodoo. Necessarily so, as slaves would have had to work hard to gain any influence or control over their lives, magical or not. More than this, slave-owners worked tirelessly and ruthlessly to stamp out non-Christian beliefs, especially in the wake of the slave rebellions in Haiti in the late eighteenth century.[23]

Between 1769 and 1803, French Louisiana fell under Spanish rule and, as a result, gained in importance and wealth. In this period, the Spanish began to draw in their own slave populations from West Africa, through a variety of routes, including some smuggled through the British Atlantic colonies. The 1788 census counted 39,410 people in lower Louisiana, of which 20,673 were classed as slaves: that is, about half of all those living in the area – but, taking into account the Natchez, this meant that slaves outnumbered free (white) people by about three to one. About a quarter of all slaves were from the Congo region. It would be this population that would give Congo Square its name (see Figure 2.12).

Figure 2.12
Congo Square offers few clues to its past. (Photograph taken in October 2001.)

the magic city 75

Congo Square was sited outside the walled city of New Orleans, just to the north of what is now Rampart Street (where the ramparts of the city used to be). During the eighteenth century, it was a market-place, where slaves gathered on Saturday afternoons and Sundays to trade (permitted under the colony's in/famous *Code Noir*, which had a provision – usually ignored – that prevented slave labour on Sundays and holidays). With the colony unable to feed itself, it was essential for the slave economy that slaves were able to subsist on the basis of their own labour: exchange of foods, medicines and other domestic goods became a vital part of this. This market was remarkable not only for its exchanges between African slaves but also with local Native American bands, including the Natchez. Congo Square was a unique site – a threshold space – where an extraordinary cultural and economic exchange, and syncretism, took place.

In 1803, the Spanish handed Louisiana back to the French but, within 21 days, the French had sold Louisiana to the United States. From 1804 onwards, New Orleans was not only taking in slave-owners and their slaves – whether refugees from the slave rebellion in Haiti in 1791[24] and/or from later expulsions from Cuba in 1809 – but there was also an influx of Americans: that is, white European Americans. French, Spanish, African and now Anglo cultures co-existed, uneasily, in New Orleans. Johnson puts it like this:

> While the city, during its long colonial history, had accommodated a number of other incoming cultural groups – the Indians, the Africans, some Germans, and a few Spanish – it had never had to deal with anything like the numbers, assertiveness, determination, or sheer foreignness represented by the American invasion. (1991, page 19)

The first American governor of New Orleans, William Claiborne, removed the ramparts of the fortified city, thereby unintentionally both enlarging the *Place des Nègres* (Congo Square) and also making it more accessible from the 'French' quarter. Though it was renamed *Place Publique*, in the 1810s the area was mainly associated with the Congo Circus, which ironically only admitted whites. Nonetheless, it remained a site where Africans gathered and dances were performed, one in particular was known as the Congo. By the late 1870s, Congo Square was renowned for the African component of its history. Further, during the Cotton Exposition of 1884–85, New Orleans become acutely aware of its difference from other American cities, one of these differences being that it was not only a French city, but an African one too, with Congo Square at its heart. Johnson describes it this way:

> The city's free people of color, while they had skins of all hues and shades, remained overwhelmingly French speaking, Catholic Creoles. Most made their livings as day labourers, dock and construction workers, washerwomen, and servants; some as skilled street vendors, caterers, nurses, midwives, seamstresses, hairdressers, barbers, and musicians; and a few as businessmen, money lenders, and planters. [They] also established the city's major free-black schools, benevolent associations, social clubs, and literary and musical societies. Consequently, Congo Square's Sunday crowds came virtually entirely from the city's Creole community [...] If Congo Square attracted the lower orders from all over New Orleans to its coarse carnival, circus, and bearbaiting spectacles, it also served as the Faubourg Tremé's [as the bordering neighbourhood was known] public meeting ground. (1991, page 35)

Although the crowds were predominantly French-speaking Creoles, the dances and spectacles of the Square attracted much interest from the white élite, most of whom were both worried and indignant. Benjamin Latrobe was one such observer. He wrote eye-witness accounts of the gatherings of 500 to 600 'unsupervised' people, in which people would dance in the open to the unusual rhythms of African drums (see Figure 2.13).

The frenzy of the dancing and the sounds and rhythms of the music were, to Latrobe's mind, 'wild and savage'. The dancers not only moved in unfamiliar fashion, they were also barely clad. Occasionally, fights would break out between groups who showed different

Figure 2.13 'The Bamboula' dance in Congo Square, 1886.
Source: The Historic New Orleans Collection.

African tribal allegiances. West African cultures – and antagonisms – had a small foothold in America. Moreover, the dancing and the music were also associated with Voodoo rituals: through Voodoo routes, different traditions were being brought together and produced anew. Indeed, it is the peculiar syncretism of these specific beliefs and practices – arguably through the space-work of Congo Square – that *makes* New Orleans' Voodoo (as opposed to Haiti's Vodou or Benin's Voudon). As Brolin says:

> Louisiana possessed a unique culture, a unique environment. In this very different world emerged a unique brand of Voodoo – less organized than the Haitian model, influenced by the mysteries of Catholicism and the basic beliefs of European super-stition, and business oriented. (1990, pages 12–13)

Voodoo rites were usually conducted in secret locations at the edge of Bayou St John (see Figure 2.14). Nonetheless, in the 1870s, a particular 'Voodoo queen', Marie Laveau, began to gain prominence in New Orleans. As a hairdresser and businesswoman, she would have been among those crowded into the Congo. Eventually, she would even preside over some of the dances, though no Voodoo rites took place at the Square. It is very well to argue that the black and Voodoo Atlantics converge on Congo Square in the eighteenth century, but the point I am making is that, in this unique site, Voodoo was created out of a flexible

Figure 2.14 The Voodoo Dance, 1886.
Source: The Historic New Orleans Collection.

syncretism of a wide variety of beliefs. These do not begin in Africa or Europe and end in New Orleans, but circulate around the Voodoo Atlantic. Indeed, Voodoo's circulation continues, as it has gained greater and greater respectability and influence over time. In part, this is to do with Marie Laveau, but it is also about its adaptability: something it learnt in the desperate situation of the slaves, but something it also gained from New Orleans.

As we will see, Congo Square is still associated with Voodoo, but it is the persistence of the dancing and music of Congo Square that are its most widespread legacy. As Johnson puts it:

> the dance forms that emerged in Congo Square were as uniquely a New Orleans product as was jazz. And those dances almost certainly made their way into the greater American culture via the same route New Orleans jazz did. And they are today, in their various derivative forms – popular and serious – as much a part of American and, indeed, world culture as are jazz and its many derivative musical forms and influences. (1991, page 43)[25]

Congo Square was a passionate and desperate site where African slave culture was drawn in, exhibited, produced, maintained and circulated. In some ways, that culture – in its Voodoo respects, at least – crystallised in the in/famous figure(s) of Marie Laveau.[26] According to Brolin (1990, page 14), slave-owners had long made use of African forms of traditional (homeopathic) medicine to 'take care' of their slaves: that is, protect their investments at no (or little) cost to themselves. It was, indeed, common practice among slave-owners to use Voodoo medicines and charms (see Figure 2.15).

In this context, Voodoo would gain a certain currency both in the African worlds of New Orleans and in the white worlds too. This double economy was also enabled by the syncretism of African and Catholic spirits and saints. Voodoo, in its iconography and

Figure 2.15
A planter and Voodoo charm, 1886.
Source: The Historic New Orleans Collection.

cosmology, could be presented as based in Christianity: using the staple Christian magics of prayer, saintly spirit guides, visions, spirit possessions and talismans.[27] Within Voodoo, there lay the potential for it to work in the threshold spaces between the African and European worlds. However, after the American Civil War (1861–65), Voodoo became more organised, more tightly-knit and anti-white in sentiment. At this time, a number of secret Voodoo societies formed (Brolin, 1990, page 18). The societies were ruled by kings and queens and practised Voodoo ceremonies: it was the queen, however, who took centre stage at these ceremonies. The most significant of these queens was, and still is, Marie Laveau (1796?–1881) (see Figure 2.16).

Perhaps Laveau's talent for the theatrical, her business acumen, and her personal charisma together with her knowledge of personal secrets acquired during her hairdressing days helped propel her to prominence. In any case, Laveau's reputation eclipsed that of other queens (it is alleged that this was accomplished by a combination of brute force and the strategic use of gris-gris) making her an influential figure on the New Orleans scene. In the process, Laveau put her unique stamp on Voodoo. Claiming her followers were Christian, Marie Laveau added statues of saints, prayers, incense, and holy water to the traditional Voodoo rites which had for some time incorporated snakes, a black cat, roosters, blood drinking, and fornication. (Brolin, 1990, page 21)

Figure 2.16
Marie Laveau and her daughter, c. 1881.
Source: The Historic New Orleans Collection.

Marie Laveau built a formidable reputation in both the African and European worlds of New Orleans. She also converted this reputation into money and was able to build herself a house near Milneburg (Pontchartrain Park), while living at a house on St Anne Street in the French Quarter. From her home, she carried out a successful Voodoo enterprise: selling charms (gris-gris), removing curses, fortune-telling and the like. Occasionally, she would conduct ceremonies outside St Louis Cathedral, the focus of the Catholic community, in Jackson Square (see Figure 2.17).

Figure 2.17
St Louis Cathedral and Jackson Square, New Orleans.

By the time of her death, Marie Laveau had created a popular interest in Voodoo, with eve of Saints' day celebrations (such as St John's and All Saints Day[28]) drawing large crowds of worshippers. These dates united – syncretised – the wide variety of belief systems in New Orleans, whether European Christian, African Animism, Haitian Vodou or New Orleans Voodoo. Though never as significant, Marie Laveau's daughter – also known as Marie Laveau – continued the female-headed family business. Since the Marie Laveaus, Voodoo has suffered ups and downs. In contemporary New Orleans, however, certain legacies are highly visible, though some are less obvious. The most visible of the legacies relate to the tourist trade, as you might expect. New Orleans has a reputation for its Voodoo, ghost and vampire tours (something I will return to in section 3.7).[29] For now, it is worth noting that Marie Laveau's tomb is a popular stop on many tours (see Figure 2.18).

Figure 2.18
Marie Laveau's tomb in St Louis cemetery No. 1, New Orleans.

What interests me is the marking of the tomb. The triple XXX mark is part of a rite invoking a wish (much like blowing out candles on a birthday cake). Around the tomb, many people leave objects and offerings. These offerings, as Freud would observe, are closely associated (by similarity and condensation, contiguity and displacement) with the wish whose fulfilment is being requested. People visiting the tomb regularly whisper wishes under their breath, even if they don't partake in some

Figure 2.19 Occult practices in Jackson Square.

Figure 2.20 Witches out to play on Halloween.

Voodoo-related rite. After all, you just never know.

In many ways, you get a sense that magic and witchcraft can appear at any time in New Orleans. Jackson Square remains a place where fortune-tellers gather (see Figure 2.19). Each fortune-teller presents their own special brand of fortune-telling. They mostly offer palm and tarot readings, but one or two base their 'unique selling point' on the reputation of Voodoo. All the while the Catholic Cathedral looks quietly on. Meanwhile, there are a wide variety of Voodoo-themed shops in New Orleans and a Voodoo Museum.[30] Every year, Halloween affords New Orleans an opportunity to openly celebrate the occult (see Figure 2.20). The phantasmagoria of magic appears to be everywhere in New Orleans.

On North Rampart, there is the Voodoo Spiritual Temple. The Temple was founded by Priestess Miriam and Priest Oswan Chamani. Although born in Mississippi, Priestess Miriam began practising Voodoo in various spiritual churches in Chicago. It was here that she met Priest Oswan, who was an 'obeah man' from Belize. Together, in 1990, they founded the New Orleans Voodoo Spiritual Temple. Oswan died in 1995, so Priestess Miriam continues to run the Temple and serve its congregation alone. For Priestess Miriam, Voodoo is a means to get the spirits (or *loas*) to answer people's prayers. To do this, Priestess Miriam offers a wide variety of Voodoo services, from bone, palm and card readings to rituals, such as weddings and snake dances, to potions and gris-gris (for purposes such as protection, success and love) and traditional medicine. In the Temple, there are various shrines at which rituals and readings take place. The shrines themselves embody the syncretism of Voodoo, with a mixture of icons from Catholicism and African Animism as well as various offerings to the spirits and more earthly items (see Figure 2.21). Some shrines directly resemble those to various saints and the Virgin Mary in the Catholic Cathedral (see Figure 2.22).

Figure 2.21
A shrine in the Voodoo Spiritual Temple.

Figure 2.22
The Virgin Mary shrine in St Louis Cathedral.

Figure 2.23
A Voodoo chalk mark in Congo Square on All Saints Day, 2001.

Two of the key dates in the Voodoo calendar are Halloween (31 October) and All Saints Day (1 November). In 2001, Priestess Miriam and the congregation, including musicians and dancers, celebrated Halloween with a midnight ceremony in Congo Square. The following day, evidence of this and other Voodoo festivities were chalked on the ground (see Figure 2.23). On All Saints Day, Priestess Miriam organised a sidewalk parade through the French Quarter (see Figure 2.24). She led a small group of worshippers (which also included two cultural geographers and an anthropologist[31]) from the Temple, down Dumaine, to Jackson Square. Some of the congregation had just dropped in, but others had travelled large distances just to attend the various ceremonies, for example from Chicago and New York. Indeed, some of the worshippers were regular visitors, despite belonging to Voodoo congregations in their own cities. Before the sidewalk ceremony, each person was given a flag or something to carry.

real cities

As we proceeded down the street, we would stop at each corner and Priestess Miriam would pour rum on to the ground, while rice and sweets were scattered, as offerings to the *loas*. Crossroads have a great significance in Voodoo beliefs, both as a site where choices are made, but also where energies and movements come together: a place of both joining and parting. Indeed, one of the key figures in the Voodoo pantheon is Papa Legba (commonly associated with Saint Peter), who is the keeper of the gates and crossroads. It might be observed (following Benjamin above), at this point, that gates and cross-roads are exactly the kinds of threshold spaces where magic might emerge.

Figure 2.24 Priestess Miriam, a New Orleans Voodoo priestess.

Eventually, the procession arrived at Jackson Square. Priestess Miriam drew a wide circle in rum on the pavement directly in front of St Louis Cathedral. In the circle, Priestess Miriam, with the musicians and dancers, began to observe Voodoo prayers and rites, while outside the circle the rest of us danced to the beat of the drums (see Figure 2.25). Rice and sweets were scattered about. The spectacle attracted some attention from passers-by, some of whom passed hostile comment at the seemingly

Figure 2.25 A Voodoo ceremony in Jackson Square, All Saints Day 2001.

irreligious and irreverent proceedings. A few, however, joined in, including one dressed convincingly as a vampire.

Two things interest me about this Voodoo ritual. First, is the marking of space: the city streets are marked as sacred spaces, where spirits come together and respects are paid. An alternative spiritual geography is traced out by the temporary space of the procession. The day after, little trace remained that the ceremony had ever taken place. Magic in cities can quite often be highly localised in space and time. Second, is the flexibility of Voodoo, both in drawing people into its circle, and also in using and reworking spaces. This is so both for the ceremonies conducted in public spaces, at cross-roads and outside the Cathedral, and also for those in the private grounds and in the Temple's rooms. All, however temporarily, are threshold spaces in which magic can emerge.

Voodoo can tell us about the creation and maintenance of beliefs (and not just magic) in cities. More, it is emblematic of the phantasmagoric circulation of magic through city life: not just of a Voodoo Atlantic that circulates through West Africa, the Caribbean islands and American cities, but also of a phantasmagoria of magic that sees Voodoo appearing in a variety of guises in the city. It is worth noting, then, that the Voodoo Spiritual Temple now has a sister organisation in Russia. Voodoo is still a creative force, acting as a crossing point for energies and movements, while New Orleans itself is a unique site in these circulations. Voodoo's popularity has, if anything, gained over the last two decades. Indeed, its charms, effigies and potions seem to promise that one's wishes can be fulfilled. For many, these wishes revolve around the desire for wealth and money. In the next section, we will see how the desire for wealth, the magic of money and Voodoo practices can combine to create a Voodoo economics at the heart of that most rational and calculative of enterprises, the capitalist business.

2.5 Stick it to Them: cities, corporations and Voodoo economics

In this section, I would like to explore in a little depth the idea that magical influence might be used in the economy. In particular, I will look at how money and Voodoo might be used to fulfil economic wishes. The idea that the economy itself might be somehow magical is not new. The mysterious ways of money and financing have often been called 'Voodoo economics'. By ascribing Voodoo status to economics, the intention is to suggest that the management of the economy is somehow both fake and corrupt. It also suggests a conscious effort to control the economy through deceitful and malevolent means. The term was most often applied during the Reagan era, for example, to describe the way that debt was being used, by occult means, to create a false and dangerous economic world. The argument was that this magical sleight of hand could not continue. For David Harvey, this 'voodoo economics' was having real effects on urban form, which he called 'voodoo cities' (1988). $35 billion of debt had financed a whole series of high-profile building projects across American cities. For Harvey, these projects were unsustainable: the trick would eventually fail. Magic and the economy, clearly, *should* have nothing to do with one another (as Park might have wished), especially if Voodoo economics was to have real effects on the urban landscape.

Money itself – and its accumulation – is commonly assumed to have magical influence.[32] At this point, I would like to turn to Simmel's discussion of money's capacity to fulfil wishes (Simmel, 1900). Here, I am interested in how money might have emergent (magical) effects based solely on its accumulation. In a discussion of the 'psychological mania for accumulation', Simmel identifies a trait in people that resembles hoarding by hamsters (1900, page 239). The miser, he says, gathers up money, but refuses to put it into circulation. This miserly behaviour, he argues, is properly described as greed, a kind of primitive accumulation. Moreover, the miser

> consciously forgoes the use of money as a means towards any specific enjoyment, he places money at an unbridgeable distance from his subjectivity, a distance that he nevertheless constantly attempts to overcome through the awareness of his ownership. (1900, page 242)

The miser exhibits an intense ambivalence towards money: it must be kept close and remote, at one and the same time. In many ways, it works in the same way as a taboo: both articulating what is desired and what is feared. The miser is caught in a trap set by what he wishes for: money. The stockpiling of money appears, to the miser, to confer on him (or her) social power and the freedom to act, even if conditions are uncertain. In other words, the possession of money appears to confer a certainty of satisfaction. However, money does not always give you what you wish for. Simmel puts it this way:

> The subjective consequences of a fulfilled wish are not always the exact complement of the state of deprivation that originally brought about the wish. The desire for an object is not like a hole that is filled by the possession so that everything remains as it was before the wish [...] The trivial wisdom that the possession of something we wanted usually disappoints us is, for better or worse, correct and we become conscious of this otherness in possession only as a fact unaccompanied by any feelings [...] If our wish does not extend beyond money towards a concrete goal, then a deadly disappointment must follow. (1900, page 243)

This is to say that, if the miser's wish is merely to possess money, then their wish for happiness that stockpiles of money embodies will be thwarted. Nevertheless, money does not lose its charm. It continues to provide the sole focus of interest for people. Even so, Simmel argues that moderns have moved beyond this miserly sensibility. For him, moderns no longer have a belief in the magical influence (social power) of simply stockpiling money. Instead, they are governed by the fantastic idea that money might be used to enable the exchange of all 'things'. For Simmel, this creates a flattening of the difference between things:

> where money is available in huge quantities and changes owners easily. The more money becomes the sole centre of interest, the more one discovers that honour and conviction, talent and virtue, beauty and salvation of the soul, are exchanged against money and so the more a mocking and frivolous attitude will develop in relation to these higher values that are for sale for the same kind of value as groceries, and that also command a 'market price'. (1900, page 256)

Salvation for people's desires and sentiments appears to lie in the fact that money makes all commodities available for exchange. Money is the intermediary between people's sentiments and desires and the satisfaction of those desires and sentiments in commodities (1900, page 80). But there is no satisfaction to be had in these commodities, as money gives people a sense that all things are exchangeable against one standard, that of money itself. This makes people blasé about the differences between commodities, as they now all appear phantasmatic, colourless (1900, page 157). Magically, money offers satisfaction with one hand and takes it away with the other.

It is of interest, therefore, that Voodoo deities sometimes demand money and that Voodoo shrines are often bedecked with coins and notes of small denominations. In Voodoo, 'exchange' involving money is asymmetrical: money is offered freely, but the deities are capricious and there is no reciprocity in the act of offering money. The magic of money, in Voodoo, remains meaningful. Perhaps because of this, Carayol and Firth, authors of *Corporate Voodoo* (2001), became interested in the idea that Voodoo might contain within it an abstract belief system capable of determining business practices. For them, Voodoo is a set of basic principles of action, whose magical influence can improve business practices. Ironically, then, Voodoo is co-opted as one of those calculative and scientific rationalities that Park thought so modern, so urban.

For Carayol and Firth, the world 'is crying out for Magic' (2001, page viii). They argue that what the muggle[33] (non-magic) world needs is Voodoo. Their wish is to initiate businesspeople in the ways of Voodoo. They compare and contrast normal business practices with that of Voodoo. Where normal business believes in hierarchies, conformity, micro-management and control, Voodoo offers a kind of wild liberation, more in tune with the African rhythms of the drum and the wild dancing of its rituals. Recognising that Voodoo is a religion founded in the barbarism of slavery, Carayol and Firth suggest that Voodoo has a double consciousness: 'Voodoo inhabits harsh reality and Voodoo inhabits the dreamworld' (2001, page 4).

Voodoo, then, much as dreams were in the last chapter, is a way of liberating people from the ordinary constraints of dominant business mind-sets: it is a way to conduct business on the basis of both new principles and a better accommodation with the earth and nature. In this sense, Voodoo is less about inflicting harm than about expressing personal joy. In *Corporate Voodoo*, Carayol and Firth side with (what used to be called) the New Economy: that is, with internet and communications companies. For them, Voodoo is on the side of fast, speeded-up business. Through studies of companies such as Virgin, Vodafone and Tesco, they show how a fast, risk-taking entrepreneurial spirit can produce magical results. From these case studies, they produce a series of principles on how to conduct Voodoo corporate business: for example, 'voodoo connects to a powerful future', 'voodoo accelerates wealth and education', 'voodoo believes differences are creative', 'voodoo is the only means necessary', 'voodoo is more than mere magic, it's your magic', and so on (summarised on pages 196–205). Thus,

> Initiates of Corporate Voodoo create Fast business [...] But Muggles are not the sad, sleepy, harmless giants you might imagine them to be. Here's the secret: even Muggle organisations already practise their own Corporate Voodoo [...] The Corporate Voodoo practised by Muggles is a spell, a curse, it's what keeps people there, what neotonises them, what makes them into zombies. It is what keeps people in check, keeps them faintly motivated but mostly disillusioned and hopeless. It is what breeds powerlessness. (2001, pages 168–169)

There are echoes in this of the idea that people sleepwalk – or act like zombies – on their way through modernity. But, correctly practised, Corporate Voodoo can awaken people, enabling them to realise their own personal dreams. Using Corporate Voodoo, you can achieve your business goals. (Echoes of Eric Cartman here: see section 1.2.) Though Carayol and Firth argue that Corporate Voodoo is a powerful means of raising the curse of zombie corporate culture, there is almost nothing on the magical practices of Voodoo in the book – there's no 'how to use Voodoo dolls against competitors' section, sadly! What is of interest is, first, the presumed magic of the (new) economy and, second, an intense ambivalence towards Voodoo outcomes: on the one hand, it produces hopeless powerless zombie workers; and, on the other hand, joyous successful powerful capitalists.

Corporate Voodoo suggests that there is – or can be – magical influence in the economy, whether the new economy or the urban economy. And, whether through spells or a belief system, that the economy can be influenced by magic. However, there is also something behind all this that implies that money itself has magical influence. One important aspect of this is the supposed capacity of money to fulfil wishes: of investments to produce large returns (we can see this in the presumed charmed life of dotcom companies), or for huge companies to grow from small investments (e.g. Orange or Microsoft): make a wish and throw a coin in the well. Significant here is the assumption that there are always sectors of the economy that will grow magically quickly – just like the beanstalk from Jack's bean. Part of the phantasmagoria of magic is the endless procession of 'magic beans' through the city. But, as Simmel warns, money – as with much magic – sets you up for bitter disappointment.

For Simmel as much as Carayol and Firth, there is an intense ambivalence towards the magic of money. For Carayol and Firth, Voodoo is both cause of and cure for corporate culture's zombie-like behaviour. Simmel, meanwhile, refuses to see money as either on the side of magic or on the side of abstract rationality: its unique character lies in its capacity to be on both sides at once. In Voodoo economics, there is the possibility both that wishes will be thwarted and also that the path to wealth and prosperity might be opened up.

The bitter disappointments of neither magic nor money have stopped people attempting to influence their environments – the urban landscape – through magic and money. There are many practices through which wealth and prosperity, and more harmonious ways of living, have been sought. In the next section, we will look at *feng shui* – the art of placement. *Feng shui* is illuminating both because it involves the direct manipulation of space and the distribution of objects in space (i.e. it participates directly in space-work) and also because it gives us an deeper sense of the (occult) systems of energy that might lie beyond the visible worlds of the city.

2.6 Singapore and the Fine Art of Placement: building fortunes in the city

Cities are magical places. Magical beliefs and practices arrive in cities through many (occult) routes and, once there, these can be animated and reanimated. Far from being in opposition to calculative and scientific rationalities of the city, magical belief and practices seem very much a part of them. Magic seems to be woven into the fabric of city life. In this section, we will discover that magic is also bound up in the production of the physical spaces of the city. An example of this is *feng shui* (usually translated as 'wind and water'). *Feng shui* is more than just evidence that magical rationalities are alive and well in the modern city, as it also entails a perspective on the nature of materiality and the system of energetics within which cities exist. Nevertheless, *feng shui* is also confirmation that beliefs in the supernatural, or occult, dimensions of the city underlie practices designed to make the urban landscape more favourable: for wealth and prosperity, for harmony and balance. *Feng shui*, from this perspective, tells us about the ordinary magics of space-work in cities: that is, the ordinary magics built into the production of urban space, designed for wealth and prosperity.

Feng shui, the Chinese system for understanding how the placement of things can influence people's lives, is widely known. There are many books offering advice on how to redesign interior living spaces, both in the home and at work, to produce the most auspicious arrangements. For Evelyn Lip,

> The practice of *feng shui* requires one to learn to live within the natural environment and to respect the forces in nature. The rudiments of *feng shui* spring from the timeless wisdom of natural science and geography. They can be used as a vehicle to achieve a better understanding of the natural environment, to develop a deeper respect for culture and to design a building more conducive to the habitation of man. They are applicable to different global cultures and it is timely that *feng shui* has finally become an universal practice. (1997, page 62)

Feng shui may not be universal, but buildings designed on its principles can be found outside China in New York, San Francisco, London and Los Angeles. In Singapore, for example, many buildings have *feng shui* principles designed into them (Noble, 1996). An example is Suntec City's 'The Fountain of Wealth', which is claimed to be the largest fountain in the world (see Figure 2.26).

According to a publicity leaflet, the fountain 'radiates with strong positive energy caused by the presence of powerful negative ions'. The circulation of water through the fountain echoes the circulation of capital through the nearby buildings. Indeed,

Figure 2.26 Suntec City's propitious 'The Fountain of Wealth'.

putting a right hand in the water and circling the fountain three times clockwise is said to be lucky, enhancing not only one's fortune but also one's well-being. More than this, the fountain 'symbolises unending completeness and unity', a unity that underpins the financial investment in Suntec City. Other examples of *feng shui* design in Singapore include some of the department stores on Orchard Road, such as Ngee Ann City and Wheelock Place.

By placing things according to certain principles, it is believed that the energies ('chi') of the world can be influenced to create better – or worse – conditions for habitation (i.e. *living in*). Lillian Too explains:

> Think of this 'magic' as energy, life force or chi that mutates and transforms, creating extraordinary changes both within individuals and in the environment around them. Imagine being able to capture, store, accumulate, and manipulate this chi into powerful, luminous energy that illuminates and improves every aspect of life. (2001, page viii)

Feng shui is part of a wider set of Taoist beliefs. In this system, there is a basic belief that the universe is made from two elements: one called *yin*, the other *yang*. In this cosmology, everything can be classified as either a *yin* or a *yang* element. Thus, the earth and moon are *yin*, while the heavens and sun are *yang*. *Yin* and *yang* are also gendered female and male. If there is imbalance between these elements in any living creature's life, then it can be harmful for their happiness and health. The skill of *feng shui* is, therefore, to bring these elements into balance. However, balance is not simply achieved by a 50:50 mix of *yin* and *yang*: Lip shows that the very fabric of space must be considered. Design must take into account

> the physical land form, climatic conditions, geographic location, and so on. For example, in Beijing where the cold, dusty winds come from the north, it is advantageous to have the windows and doors of buildings facing south with solid north-facing walls to avoid the dust and cold. (1997, page 11)

This may seem like common sense, but it is important to note that this particular magic is ordinary and practical (a bit like Voodoo, but with a better reputation). The space-work of *feng shui* also includes a philosophy of materials and forms. There are five elements: wood, fire, earth, gold and water. Each of these elements has associated with it orientations, seasons, climatic conditions, colours, animals, parts of the face, human qualities and parts of the body. Thus, wood is associated with east, spring/summer, wind, green, goats, eyes, anger and muscle; while earth is associated with centre, autumn, moisture, yellow, ox, nose, thinking and flesh. Each aspect of each element (such as eye or nose) is made up of either *yin* or *yang*. The art of placement must assess these elements and their qualities, balancing the *yin* and *yang* by combining the elements in particular ways. Divining the correct placement of things is an extremely fine art: failing on one element would throw the whole arrangement out of balance, like getting a piece wrong in a jigsaw puzzle.

To determine the appropriate placement of things, it is interesting to note the development of

> two schools of *feng shui*: one makes use of instruments and methods of calculation, while the other depends on assessment of land form and acquired experience. (Lip, 1997, page 20)

In some ways, this echoes the split between science and magic in western thought. On the calculative side, Too (2001) shows how the orientations of buildings, the sectors (using a spatial grid) within buildings and the placement of things within sectors can be given a numerical value. This numerical value can then be used to assess the balance of things within any space, at whatever spatial scale (within a room, house, neighbourhood, city). On the more experiential side, elements – such as orientations – have favourable and unfavourable positionings. It is experience that allows the correct placement of things, not the science. The magic of space, indeed, might emerge from unexpected or intuitive positionings of things.

Figure 2.27
Canary Wharf's auspicious waterfront, as it was in the early 1990s.

Using this system, it is also possible to read out of buildings their various elements and, then, to determine what might need to be done to improve their siting within the energetics of the universe. In other words, *feng shui* is not just a means of designing, of doing space-work, but also a diagnostic capable of interpreting the existing influence of things in the world. For example, each element is associated with specific shapes: wood with rectangles; fire with triangles; earth with squares; gold with round shapes; and water with flowing lines. Thus, one can take a building such as Canary Wharf (Figure 2.27) and judge it against its rectangular (wood) and triangular (fire) elements, its orientation and the nature of water that surrounds it to assess whether it is favourably or unfavourably placed for habitation.

In principle, *feng shui* suggests that only five pairings of elements are favourable: water with wood; wood with fire; fire with earth; earth with gold; gold with water. Meanwhile, each element should take specific forms. Thus, water near a building must be deep if it is the sea, be calm if it is a lake, and clear if it is a pond. And so on. On this basis, Canary Wharf does quite well, with the right combination of elements and surrounding water. However, its entrance faces towards the west and away from the water. This is, be advised, less auspicious.

Cities and buildings have both been designed according to *feng shui* principles. This is particularly so in Chinese cities (such as the Forbidden City in Beijing) or cities with strong Chinese influences, such as Singapore. Moreover, *feng shui* can also act as a diagnostic of habitation, favourable or unfavourable. The impulses behind the use of *feng shui* are utopian, acting on the desire to create a city as favourable as possible for the creation of wealth and prosperity, harmony and balance, for its inhabitants. But the circulation of magical beliefs and practices through cities is not always intended to benefit everyone. Indeed, the phantasmagoria of magic provides opportunities for some people to hide the harm they do, whether by moving their practices through occult globalisations, or by hiding what they do in the shadows of other places. The consequences of the globalisation of magical beliefs and practices can be deadly serious.

2.7 London and the Global Trade in Magic: the high cost of healing

On 21 September 2001, the torso of a young boy, whom detectives would call 'Adam', was found floating in the River Thames, London. The autopsy discovered that Adam had probably been killed by trauma to the neck. His limbs and head had been removed after he was killed. Detailed forensic analysis of the contents of the boy's stomach found items – pollen and food – that indicated that the boy had only recently arrived in London and was probably from Africa. More alarmingly, in December, detectives disclosed that quartz, clay pellets, rough gold and bone fragments in the boy's lower gut seemed to be some kind of 'magic potion' and therefore, they said, the boy was probably the victim of a ritual killing. The magic potion, they added, also contained some unidentified vegetable matter.

The British police expressed incomprehension at the culture and beliefs that might lie behind such a brutal act. British newspapers, meanwhile, shared the detectives' horror over Adam's murder (see, for example, Vasager, 2002). The shared assumption was that this was a muti murder, associated with South African 'black' magic. There were hints that black South Africans had smuggled a young boy to London for the simple purpose of sacrificing him for their own personal gain. So, on 14 April 2002, the detectives travelled to Johannesburg to talk to police with experience of dealing with ritual murder and other witchcraft-related crimes. The British detectives also consulted other experts in muti and muti murder and they toured the muti markets of Johannesburg, such as the Mai Mai market (Figure 2.28) and Faraday Centre (Figure 2.29).

Figure 2.28
A muti shop at the Mai Mai market, Johannesburg.

The assumed link to South Africa was probably not just made on the basis of Adam's 'African' stomach contents. The British detectives would have been aware both of the witch murders (both by and of witches) in South Africa's Northern Province and also of Johannesburg's reputation as a leading centre in the world-wide muti trade.[34] Indeed, South African

Figure 2.29
The Faraday Centre muti market, under one of Johannesburg's freeways.

Figure 2.30
Animal muti at the Faraday Centre.

Figure 2.31
Plant muti at the Faraday Centre.

police occasionally raid and arrest muti traders on suspicion of selling human body parts. For example, in May 2002, a Soweto woman had to go into hiding after being accused of selling a human eye at the Faraday market. However, the human body parts usually turn out to have belonged to chimpanzees or other primates. Not that the fact that monkeys are killed for sale as human body parts is in any way reassuring, of course. As the British police walked around Johannesburg's muti markets, they would have seen the body parts of many animals (Figure 2.30) and also a huge variety of herbs, bark and other plant extracts (Figure 2.31).

The vast majority of the muti trade is for medical purposes. With about half of South Africans being classed as 'poor', and with 95% of these being classed as 'African' (i.e. black), muti is a vital part of health care for millions of people. It therefore performs one of the classic functions of magic: it provides protection and curatives, cheaply and homeopathically (see Ashforth, 2001). In fact, the muti trade is vast and forms a core part of the informal economic sector in South Africa. In her research, Haripriya Rangan has argued that 'more than one million households are involved in collecting the roots, bark, bulbs, and stems of over 500 plant species' (2003, page 2), with these goods coming to Johannesburg from all over southern Africa. It is worth noting that the muti trade is primarily a women-centred economy. If muti is part of a regional economy, then it is also part of an international – and possibly global – economy too.

While in South Africa, it is important to note that the British detectives also consulted Credo Mutwa, a widely-known and respected *sangoma*.[35] Hearing the details of the case, Credo confirmed that the boy had been part of a sacrificial ritual. In the ritual, he explained, the boy's blood would probably have been drunk from his skull. The fingers of the boy would be used in charms, while his bones would be crushed into a paste to give people strength. It was likely, Credo added, that the sacrifice would have something to do with a sea or water goddess (with the body being found in the river Thames). Finally, the boy was found in orange shorts which, for Credo, suggested 'resurrection', so he deduced that it was likely that Adam's killers were close relatives. Credo Mutwa concluded that the ritual – and indeed the body of the boy – was not South African in nature, but probably West African, and very probably Nigerian.[36] Nevertheless, the detectives returned to England, they believed, without a clear lead. And, on 19 April 2002, Nelson Mandela made an impassioned appeal to people in London and (South) Africa to come forward with information that might determine Adam's identity and his killers.

Six months after the detectives returned, the *Daily Express*, frustrated with lack of progress – a claim that both disguised and fuelled racist motivations concerning Africans in London – launched its own investigation into the Adam case. On 3 November 2002, the paper's special reporter Paul Smith published his findings. He had tracked down two 'witchdoctors' currently living and openly practising witchcraft in London. The headline was sensational:

> The witchdoctor who trades 'slaughtered children's fingers' as Voodoo sex charms. (2002, pages 18–19)

Paul Smith began his exposé:

> A Voodoo medicine man is playing his evil brand of black magic in Britain, offering the severed fingers of children as charms. A piece of flesh from a ritually slaughtered African child costs £5,000 and is turned into a gruesome pendant to give the wearer 'incredible sexual power'. (2002, page 18)

The *Daily Express* reporter continued:

> The number of 'African spiritualists' practising Voodoo has exploded in the past twelve months, fuelling the demand for human meat – the trade in both live children and dismembered body parts smuggled into the UK from Nigeria, Ghana and South Africa. (2002, pages 18)

Such statements remind me of Frantz Fanon's bitter observation that, through the reactions of white people, he discovered that he, as a black man, was 'battered down by tom-toms, cannibalism, intellectual deficiency, fetishism, racial defects, slave-ships' and all the other horrifying stereotypes associated with Africans (1952, page 112). Such stereotypes clearly have not gone away since Fanon was writing: to the mix, the *Daily Express* adds witchcraft, along with sadistic rituals and child murder. The report concluded, as if confirming African barbarism, that the ritual murder of Adam would have been as painful as possible to strengthen the power of the magic.

Despite the supposed explosion of Voodoo, the *Daily Express* could only track down two witchdoctors, one of which would only sell *cures* for curses. The other witchdoctor turned out to be from Nigeria: he was quoted as saying: 'If you want to destroy a person, you can use Voodoo magic, but you can also use Voodoo for love' (page 19). Overlooking the ambiguity of this statement, and also the injunction to use Voodoo for love, the *Daily Express* begins to build circumstantial evidence against this witchdoctor, aligning Yoruba rituals with the murder of Adam. The *Daily Express* also pointed to Voodoo ritual murders in Belgium and Frankfurt. No new leads resulted from their investigative journalism. Yet, despite the *Daily Express*'s attempt to make magic strange, foreign and dangerous, it is never that far away. On the page before their Voodoo exposé, the *Daily Express* carried a story about the spell-binding party held to celebrate the release of the latest Harry Potter film (page 17).

In September 2002, police revealed that forensic examination of the 'magic potion' in Adam's stomach was continuing. The bone analysis indicated that Adam had come from an area surrounding Benin City and Ibadan in south-west Nigeria – much as Credo Mutwa had concluded nine months earlier. Meanwhile (it was later revealed), the mysterious vegetable matter had been identified as calabar bean. This rare bean is used in West African witchcraft and was probably employed to paralyse Adam prior to his sacrifice. In December 2002, Glasgow police interviewed a woman whom they believed to be a vital witness. She was subsequently deported to Nigeria. In July 2003, Dublin police arrested Sam Onogigovie, who apparently was the estranged husband of the woman in Glasgow. He is believed to be a close relative of Adam, possibly his father. Onogigovie now faced extradition to Germany in connection with similar crimes there. In dawn raids on 29 July 2003, nearly 200 police officers in London arrested 21 people, connected to Onogigovie, whom they believed were part of a child-smuggling ring. This ring, police allege, not only brought Adam to London but was also involved in widespread trafficking of people – especially children – from West Africa to Europe.

Adam's case not only shows how magic has its own kind of globalisation,[37] its own kind of urban linkages, but also the dire consequences these *occult globalisations* can have for people, especially the most vulnerable. That there is a global trade in human body parts is widely known, that there is a global trade in human beings is widely known, it is less well known that magic might have a lot to do with what and who is going where and for what purposes.

2.8 Conclusion: cities, careful wishes and the occult

My local bookshop has a signing by a witch. A flyer pushed through my letterbox offers the services of Professor Fana, who, it promises, is 'a gifted spiritual healer […] using the occult sciences and the most powerful spell [sic]'. Meanwhile a local Christian church (whose funding comes from Brazil) offers exorcisms (and has been for the last four years) to anyone who needs it. The phantasmagoria of magic can be found everywhere in cities. Even so, it is in some places more than others. By tracking magic through the city, I have been able to convey a sense both of the procession of magical beliefs and practices through cities and also of the creation of phantasmagoric sites within cities, generative sites capable of instigating new magics and new practices. Urban magics have used the city as a cosmology and a resource for developing their beliefs and practices, especially as they adapt themselves to city life.

Initially, I set this chapter up by asking whether Robert Park was right to assume that beliefs in magic are antithetical to city life. Exploring this question led in many directions – from New York to Haiti, from Haiti to New Orleans, from money to corporate management practices, from *feng shui* design principles to a fountain of wealth, from muti to murder. Each time, Park's belief that the city would triumph over magic has proved ill-founded. Indeed, quite the opposite. Not only is the city a site where magics are produced, but in it science is barely distinguishable from magic – sharing remarkably similar beliefs and practices, whether in building design or corporate principles. This is more in line with the arguments of Simmel and Freud, who were both able to see that the city, and modernity, cannot shake off its belief in magic – whether this takes form in money or a belief in the powers that lie beyond the visible. Nevertheless, Simmel and Freud believed modernity to be in opposition to magic. This more nostalgic view of magic has also proved ill-founded. The phantasmagoria of magic has thrived in cities, whether in creating new forms of magic, or in blending itself with already existing beliefs – including calculative and scientific rationalities.

Magic reveals how cities work as a site where new ideas, beliefs and practices emerge. Within cities, certain places act as nodal points through which magic flows. In allowing for crossing between different worlds, threshold spaces allow for new relationships between people and things to emerge, as Walter Benjamin observed. They can do so with great intensity. Congo Square shows that certain threshold spaces become places where new forms of magic – where new histories – can be created. Voodoo taught us, also, that these threshold spaces could include temporary spaces produced by rituals, even in the face of the Catholic Cathedral (compare Figures 2.17 and 2.25). Thus, Voodoo precisely shows how magic can operate, beyond the visible realms of the city, in the marginal spaces of the vulnerable and oppressed, under the noses of the powerful.

We can see, in all the forms of magic described above, that there is a flow of ideas in and out of the city. I have suggested, for example, that there is a Voodoo Atlantic. Others have argued that *feng shui* now has a global reach, influencing not just interior design, but the international design of buildings, including prestige projects in Singapore. The London muti murder showed that not all these circulations of magic have good – or indeed easily belittled – consequences. Out of these stories, we get a very different sense of the occult influence of magic on the city and the circulations, encounters and interactions that have direct effects on the modern city's cultural and physical forms. There is a new spatial language to describe this. Terms like arrangement, siting, positioning, orientation, contiguity, proximity all help us appreciate how the particular combination of things in cities has direct consequences for city life.

Not all ideas are going to spread and engage people in other places, but cities are sites where ideas are in constant circulation and agitation. Though its history is markedly different from Voodoo, the current popularity of magick and wiccan witchcrafts is another instance of this. In the modern city, beliefs and rituals that invoke magical worlds and seek to produce magical outcomes remain important to people. Often, it is difficult to distinguish between scientific, religious and magical beliefs. Through magic, the city life is revealed as a phantasmagoria of belief systems, all entrained by one another: people *believe* in cars and commuting, in cashpoint machines and mobile phones, in lifts and automatic doors, in temples, churches, synagogues, mosques and roadside shrines. People believe in the magical realism of modernity as much as the ordinary lottery of everyday life.

Magic speaks to those secret wishes that no longer get talked about very much: happiness, prosperity, good fortune. Even so, we should bear in mind the caveat associated with the fulfilment of wish by magical means: we have to be careful what we wish for. More than this, we have to be careful who we reveal wishes to and who pays for the wishes that get fulfilled: we must remember Adam. The city is full of wishes: wishes, needs, desires, yearnings, longings, fancies. One side the city presents to its inhabitants is fulfilment: here, you will find everything you want. Yet, as for anyone who had anything to do with a genie, the city also disappoints: it does so so thoroughly that city dwellers are barely troubled by the experience. Instead, they begin to live their lives in the midst of disappointment: the constant failure of things in cities to give people what they promise or what they want. Magic is full of promise. It offers a kind of control or influence over things that is unattainable any other way. But magic is also full of dangers. Magic, then, gives us a sense of the landscape of hardship and luck, of disappointment and hopefulness in city life.

Through the phantasmagoria of magic, city dwellers' express wishes and desires, anxieties and fears, are formed into images, into beliefs, rituals, practices, into charms, prayers and spells. Magical beliefs and practices circulate through the city, rubbing up against other beliefs, being assimilated into others, such that a generalised sense of magic might even pervade city life. As I have implied, this sense of the magical worlds that exist beyond visible realms is often associated with beliefs in such figures as vampires and ghosts. This is not the reason, however, for turning to vampires and ghosts.

In the next two chapters, instead of looking at a generalised sense of the phantasmagoric in city life, I will turn the analytical focus upon two social figures: the vampire and the ghost. Each is a figure associated with the phantasmagorias of city life, but by focusing upon them as figures it is possible to show how they themselves act as nodal points in webs of affect, meaning and power. Each creates, and is created by, very different geographies and histories of the city – the vampire is associated with the undead, the ghost is associated with the dead.

In this chapter, magic gave us a sense of the occult circulations and globalisations associated with the city. In many ways, this phantasmatic flow of magic through the city seemed quite linear, sequentially connecting one place with another. The vampire, also, is associated with flows, in particular flows of blood. Tracking blood and the vampire, however, will enable me to highlight the discontinuities and non-linearities in these flows. We have heard, also, that magic is associated with intense ambivalence, but just wait until you meet the vampire.

3
The Vampiric City
in which blood flows free

3.1 Introduction: the life-blood of the city

In this chapter, the focus is upon the vampire: a figure that appears, perhaps surprisingly commonly, in the phantasmagorias of modern city life. The vampire is an intensely ambivalent figure, lying at the intersection of often contradictory networks of affect, meaning and power. In the imagination and in popular representations, the vampire is often associated with supernatural powers, with blood and death, with undeath and immortality, with earth and castles, with aristocracy and coffins, with decadence and desire, with hunger and fear. By examining the vampire, it is possible to reveal how specific fears and desires are condensed into, or displaced on to, the blood-sucker. This is to treat the vampire as if it were a dream-like element in the phantasmagorias of modern city life. But it is also possible to track the vampire as it moves through cities and their phantasmagorias. Vampires, unlike ghosts, for example, are highly mobile in time and space, despite their reputed vulnerability to sunlight. Not only are they supposed to have super-human senses and movements, they can also — under certain conditions — move great distances or lie in wait over long periods of time. Tracking vampires shows that they have moved from the furthest reaches of urban networks to the very heart of cities. Indeed, we will discover that the city is the vampire's ideal home. If the social figure of the vampire represents something real about city life, one of the more disquieting implications is that the city is itself vampiric — sucking the life-blood of its citizens.

The figure of the vampire is most closely associated with blood-drinking. Drinking blood — whether human, animal or even that of the son of God — is far from uncommon. It does, however, provoke a reaction, for blood is associated with intense ambivalences. On the one hand, blood can represent the very stuff of life. On the other hand, blood is closely associated with death. The meanings and powers that chain themselves to blood are often highly emotionally charged: blood's purity or contamination is often a source of, or means of expression for, desire or anxiety. Nonetheless, the meanings and powers of blood are not universal, at least not in the sense that blood always and everywhere has the same meanings and same powers (see, for example, Bhabha, 1985). Perhaps what is universal about blood is that it is held to be significant, that it is thought to have certain powers or dangers. Thus, the HIV/AIDS crisis has exposed certain horrors associated with the contamination and the free-flowing circulation of blood.[1]

Freud reminds us that the intense ambivalence surrounding blood, through its associations simultaneously with life and death, can lead to it being given a vital role in social and religious rituals (1939, pages 156–157). Thus, warriors drink the (human) blood of the slain, not simply to gain their strength, but also to honour them. Vampire myths reveal other vital properties of blood: blood is heat, blood is taste, blood is life.[2] Blood is also desire. But it is not always clear what this desire is for. For Copjec (1994, Chapter 5), blood is to be understood in relation to anxiety.[3] The desire for (drinking) blood, in this view, is a flight from an anxiety.

> If vampirism makes our hearts pound, our pulses race, and our breathing come in troubled bursts, this is not because it puts us in contact with objects and persons – others – who affect us, but because it confronts us with an absence of absence – an Other – who threatens to asphyxiate us. (1994, page 128)

The vampire scares people because it embodies repressed fears and, thereby, makes people strange to themselves. In Copjec's Lacanian account, the desire for blood confronts people with an anxiety related to an absence, an absence that cannot be recognised (as such).[4] In these terms, the vampire represents an eruption of the real into the symbolic; or, alternatively, the return to consciousness of a repressed unconscious anxiety.[5] Copjec's surprising conclusion is that the unconscious anxiety associated with vampirism is the fear of the mother's breast drying up (1994, page 128). In drinking blood dry from the neck, the vampire embodies – by proximity and similarity – the child's insatiable desire for mother's milk. There is something to this. Copjec has, at least, paid attention to the dream-like quality of vampire bitings: the vampire holds the victim in hypnotic thrall, sometimes appearing (as if) in dreams, so why shouldn't both vampire and victim be dreaming of the breast?

For Ernest Jones (1931), the story is just as *uncanny*. The appearance of the vampires is based on an intense ambivalence towards the dead. The resurrection of the vampire, and its taking of life, has to do with the fear that the dead intend the living harm (a story we will pick up again in Chapter 4).[6] It is for this reason that vampires are so sadistic: their sharp fangs bite deep into the living, drawing their life slowly but inexorably from their bodies. For Jones, the rape of women by vampires also needs to be understood in this vein. The vampire, in this version, is best understood as being undead or demonic, rather than as a blood-sucker. What these psychoanalytic accounts suggest is that the figure of the vampire is overdetermined, whether as a blood-sucker or the undead. And the difference really matters, psychodynamically: to the vampire is attached feelings of fear, of love-turned-to-hate, of one guilty thirst displaced on to another, of sadism and shame.

Perhaps because of their focus on anxiety and fear, neither Copjec nor Jones consider the desire to be a vampire.[7] For both Copjec and Jones, the vampire is a figure of horror, not something that anyone would identify with, or wish to become. However, there is evidence to suggest that for some people the desire to be a vampire is all-consuming. To convey a sense of this desire, here is a story of vampire killers: killers who drink blood, who wish (thereby) to become vampires.[8] This horror story indicates exactly what is at stake in vampire tales.

On 31 January 2002, the *Islington Gazette*, London, carried the alarming front-page headline: **Vampires Shocker. Secret cult followers drink each other's blood** (Figure 3.1).

Figure 3.1
Vampire Shock Horror.
Source: Islington Gazette.

The report began:

> A secret colony of 'vampires' is operating within Islington's bars and nightclubs. Men and women obsessed with drinking human blood have met at Islington venues. Then they have allowed other vampires – or 'sanguinarians' – to drink blood from their veins. (2002, front page)

An inset picture carried a picture of Manuela Ruda making a hand gesture (resembling a Texan horn gesture). Below the picture, the caption hinted at what the *Islington Gazette* felt the real shock was:

> she was first introduced to the vampire cult in Islington. She is now facing a murder charge in Germany. (2002, front page)

Vampires? In Islington! On 1 February 2002, most British national papers carried the news that Daniel and Manuela Ruda had been convicted of murdering Frank Hackert in Witten, western Germany. The details of the case were shocking. Daniel Ruda had targeted Frank Hackert because he was a mild-mannered man and because he loved the Beatles. Frank was lured back to the Rudas' flat where he was attacked with a hammer. Then, he was stabbed 66 times in the heart. A pentagram was carved in the man's chest. Hackert's blood was collected in a bowl and the Rudas drank it. According to Allan Hall, of the popular newspaper the *Sun*, after these rituals the pair had 'frenzied sex'. The *Guardian*'s John Hooper missed this detail, but reported that:

> When the police broke into the flat they found a scalpel still embedded in his stomach with his body lying beneath a banner saying 'When Satan Lives'. They also found imitation human skulls and a coffin in which Manuela slept in the day. (Hooper, 2002, page 2)

Daniel and Manuela Ruda were convicted of Hackert's murder on 31 January 2002 and sentenced to 15 and 13 years in prison, respectively.* During their trial, the Rudas showed no remorse and instead remained defiant. The British press, however, became worried about Manuela's claims about what she had been up to while living in London. Her account, according to Hooper, raised 'the spectre of bizarre underground occult groups in Britain' (2002, page 2). It seemed as if Manuela's vampire fantasies had started during visits to Gothic-themed nights at the Electrowerkz nightclub and at special vampire nights in the (infamous) Torture Gardens sadomasochism club. Indeed, the Torture Gardens seems to be a popular place for vampires and donors to get together. There are, according to the press, many ways for London vampires to meet, including through organisations, such as Vampire Connexion, and a host of websites.[9] It seems as if the 'city of dreadful night' is actively making 'creatures of the night': indeed, that London nights are particularly sinister, full of dark practices and shadowy figures.[10]

London is not alone in having an underground vampire culture. As we will see, one of the main cities attracting vampiric attention is New Orleans, but groups have been reported in many large cities, including New York. Vampires are not necessarily drawn to cities. But, as Louis Wirth (1938) once pointed out, it is clear that cities offer people the possibility for new forms of association – and among these must be the potential for the creation of a vampire identity. More so because vampires have moved from the adult horror movie on to children's and early evening television (e.g. in shows such as *Mona the Vampire* and *Buffy the Vampire Slayer*. Figure 3.2) and in doing so they have become more familiar.

Some TV vampires are even not all that bad: Angel, for example, saves people from (their personal) demons in the *City of Angels*. Clearly, the vampire is no longer just to be feared, nor a figure of horror (despite the cold blood-drinking). In these forms, the vampire has 'moved on'. There is something alluring, enigmatic and beautiful about them in contemporary guise. Even the fact that vampires are doomed (soulless) gives them a kind of grace. Clearly, they're perfect for teenage fantasising: power and sex, death and night and blood.[11] Irrespective, and in their many

Figure 3.2
Buffy lives.

guises and disguises, vampires remain a significant cultural icon. Vampires may not have started in the city or in the suburbs, but they have certainly come to town.

For many, the vampire remains a fiend enslaved to its thirst for (human) blood. Their insatiable hunger makes vampires bestial, murderous. Terrifyingly, they can also 'turn' their victims into versions of themselves, by making them drink their blood. The vampire, then, is a thing that both kills and also offers immortal life. In both senses, the vampire is a figure of supernatural power. There is, then, something very seductive about the vampire. They are beautiful, powerful, and they can live forever. For all their strangeness, they are also very familiar: not an evil human, but super human. The puzzle is this: if there is something seductive about the vampire, some part of us that wishes to give in to the vampire – to be the vampire – then people, cities, will also have to find ways to live with the vampire.

Tracking the vampire – much as Voodoo was tracked – will allow us to see something of the city: the desires and fears embodied in strangers, racialised or otherwise; the fear of death, the desire of the living; the need to be in a contact with a time and a place; the circulation of things through cities that have very different temporalities; and the sharp bite of city life. In this chapter, I will use the figure of the vampire to move in and out of certain ideas about cities and the ways in which various worlds – the global, the imperial, the carnal – circulate and intermingle within cities.

To begin with, in **section 3.2**, I will provide a brief history of vampire plagues. On different continents, in different periods, vampire plagues sweep inwards from the very edges of Empire. The sheer quantity of stories alerts us to the significance of the vampire in imperial phantasmagorias, but these plagues also leap time and space, thereby conveying a sense of the mobility of vampires. Eventually, the vampire reaches the beating hearts of Empire, their metropolitan cores. In **section 3.3**, I will look at the various worlds that circulate through London in Bram Stoker's well-known work, *Dracula* (1897). In particular, I am interested in the role of blood and technology, since these create new (terrifying) geographies. This analysis leads, in **section 3.4**, to an exploration of how fears about blood and mortality are intensely expressed in the Victorian city, in part through cleaning blood from the streets, but also through a sense of the immortality of money in circulation. Not only money, but also desire circulates through the city. In **section 3.5**, I note that Bram Stoker's research into vampires led him to think about, and draw upon, Malay stories about the pontianak – female vampires who prey upon unsuspecting men. These myths continue to circulate through the modern city.

Women are not the only beautiful, desirable vampires. In Anne Rice's tales also, vampires are aristocratic, beautiful and doomed. In these tales, New Orleans is prominent, so this city provides the focus for sections 3.6 and 3.7. **Section 3.6** examines Anne Rice's *Interview with the Vampire* (1976), especially Louis' transformation into a vampire. Often, Louis expresses his sense of alienation. His lack of belonging is taken, by other vampires, to show his sympathy with both modernity and the modern city. **Section 3.7** shows how vampires have became part of the life of New Orleans, partly through vampire tours, partly through vampire tales: tales that draw vampirism to the very heart of the city. Finally, I conclude on the idea that cities are themselves vampiric: highlighting how cities live with vampires, but also the city's capacity to suck the life out of people.

We can start tracking the vampire in the places where they set out on their undead journeys, on the very fringes of several Empires (most now gone).

3.2 A Plague upon the West: vampire and Empire

Vampires themselves, seemingly, require very little introduction – the blood-sucking fiends have been close to our hearts for a very long time. Myths surrounding vampires have taken many different forms, in many different places, in many different periods. I don't intend to review, or even synthesise, these myths into any coherent image of the vampire. It is probably the case that variations in the form of the vampire are more interesting. The figure of the vampire is precisely shadowy, illusive; the powers of the vampire, various. But we all know they have sharp teeth and drink blood – and that they are immortal, if only to the extent that they are very hard to kill. Indeed, vampires are quite likely to return from the dead if they are not disposed of properly. The shadowy shape of the vampire is all too familiar: the black cape, the aristocratic clothing, the hypnotic eyes, and sharp white fangs. In this form, it is probably Christopher Lee as Dracula.

Dracula was, however, far from being the first vampire, whether in literature or in real life. There were vampires long before Bram Stoker wrote his classic tale (which we will turn to in the next section). To give us some sense of context, it is worth laying out a little of the history of vampirism – a context that will tell us much about blood and fear in late Victorian London: the very heart of Empire.[12] For Christopher Frayling, 'the vampire is as old as the world' (1991, page 4). He draws evidence for this from many myths, scattered across the ancient world, from Greece to Mongolia. Most cultures, Frayling argues, have stories about blood-draining or blood-drinking, whether by gods, those in the service of gods, by monsters or by humans. Key to contemporary vampire tales, he suggests, is Dr John Polidori's *The Vampyre* (1819). Nevertheless, Frayling's scouring of the various 'authorities' on vampirism throws up evidence of the diverse geographies of vampires. This is what interests me. So, in this section, I will begin with, then extend, Frayling's discussion of the vampire epidemics that swept across Eastern Europe in the late seventeenth and eighteenth centuries.

From the mid seventeenth century onwards, vast swathes of Eastern Europe had learnt to become terrified of vampire plagues, not as a fairytale, but as a fact. Frayling informs us that 'these epidemics had occurred in Istria (1672), East Prussia (1710 and 1721), Hungary (1725–30), Austrian Serbia (1725–32), East Prussia (1750), Silesia (1755), Wallachia (1756) and Russia (1771)', and that individual vampires included 'Giure Grando (Khring, Istria), Peter Plogojowitz (Kisilova, Serbia), Arnold Paole (Medvegia, near Belgrade) and the vampires of Olmutz (Silesia)' (1991, page 19). Likewise, in 1931, psychoanalyst Ernest Jones noted some of these epidemics, but also plagues in Serbia (1825) and Hungary (1832); vampire fears in Dantsui (1855) and Roumania (1899); and vampire burnings that took place in Hungary and Bucharest in 1902 and in south Transylvania in 1909.

Laurence Rickels (1999) is equally sure that vampires (really) have been among us for centuries. He cites the notorious late seventeenth-century case of Countess Elizabeth Báthory, who would suck the blood of consenting adults: she also killed her victims.[13] Fears over vampirism led to legal, medical and military investigations. Rickels found that in the 15 years after 1728, over 40 treatises on vampirism were published (1999, page 15). If there were doubts over the existence of vampires among the intellectual élite, there was actually very good evidence that vampires existed. For example, both Frayling and Rickels turn to the account provided by Dom Augustin Calmet of his investigation into vampire epidemics (Calmet, 1746). Calmet's treatise collated a variety of reports on vampires, including 'formal reports, newspaper articles, eyewitness accounts and critical pieces' (Frayling, 1991, page 92). Moreover, according to Frayling, the book became an instant best-seller.

One typical vampire tale that Calmet tells concerns Peter Plogojowitz (died 1725), who appeared to the inhabitants of the village of Kisilova, then under the control of the Austro-Hungarian Empire. Despite having been buried some ten weeks earlier, Plogojowitz managed to strangle nine people in an eight-day reign of terror. As a result of a positive identification by Plogojowitz's widow, the villagers decided that they would disinter the man's body and burn it. To do this, they needed a court order from the Emperor's officer in the region and the parish priest, the Curé. The officer and the Curé decided to appease the villagers by exhuming the body. On investigation, according to Calmet, they found that Peter Plogojowitz's body:

> exhaled no bad smell; that he looked as when alive, except the tip of the nose; that his hair and beard had grown, and instead of his nails which had fallen off, new ones had come; that under his cuticle, which appeared whitish, there was a new skin, which looked healthy, and of a natural colour; his feet and hands were whole as could be desired in a living man. They remarked also in his mouth some fresh blood, which these people believed that this vampire had sucked from the men whose death he had occasioned. (Calmet, 1746, page 101)

Calmet continues the story. The Emperor's officer and the Curé concurred with the villagers that Plogojowitz was indeed a vampire, so they thrust a stake into the breast of the body. At this point, large quantities of crimson blood spurted from the wound and from the nose and mouth of the body. As was standard practice in suspected vampire cases, the body was then placed on a funeral pyre and burnt. Indeed, this method of disposal had been used in the vampire plagues of New England. Ernest Jones (1931, page 412) reported that, in 1874, a Rhode Island man had exhumed his daughter and burnt her. According to Michael Bell (2001, page 64), her name was Ruth Ellen Rose and she was far from being an isolated case. Indeed, Bell documents over 20 vampire cases between 1793 and 1898 in the New England area. In the late eighteenth century especially, it seems that colonial America had its own vampire problem, mostly at the edges of European settlement. Vampire plagues are not solely an East European phenomenon.

In her book, *Speaking with Vampires* (2000), Luise White gives a full account of various vampire scares in colonial Kenya and Uganda. In particular, in the 1930s, it was thought that the National Fire Brigade in Nairobi was involved in vampirism. It was alleged that they kidnapped people, took them back to the fire stations and drained their blood:

> To European officials, these stories were proof of African superstition, and of disorder that superstition often caused. It was yet another groundless African belief, the details of which were not worth the recall of officials and observers. But to young Africans growing up in Kenya – or Tanganyika or Northern Rhodesia – in the 1930s, such practices were terrible but matter-of-fact events, noteworthy [...] only when proven to be false or when the details of the story required correction. (White, 2000, page 4)

For White, these stories of vampires are not simply false or irrational, but are reasonable expressions of African fears under British colonial rule. More than this, vampires are a diagnostic, revealing power relations extant in (British) colonial Africa at the time. Africans, in this period and in these contexts, did not know what Fire Brigades, medical practitioners and police did with people, but they did know they were important and powerful. Fire Brigade people dressed in overalls and wellington boots and their equipment was red. Why did they wear overalls and boots? Why was it red? What happened within those stations, whose doors were closed and which had deep pits under the fire engines? Medical practitioners were constantly draining blood from Africans: what did they use all that blood for? For White, vampirism is a reasonable assumption given the evidence to hand: more so, since there was ample real-life testimony about vampire acts by those who (claimed they) had escaped the vampiric clutches of the Fire Brigade.

Vampire rumours were rife, not because of some premodern or irrational belief in monsters, but because there were believable first-hand witnesses whose experiences could not simply be dismissed as lies or superstitions (either by sceptical Africans or by the British colonial authorities) and also because these tales 'fitted' with extant power relations and the grotesque psychodrama within the colony. Indeed, vampire stories surface in many different colonial contexts: in urban Nairobi, in rural Northern Rhodesia, in anti-tse-tse fly campaigns, in the copperbelts of Eastern Africa (White, 2000, page 6). This says much about the interweaving of commonplace fears under colonial rule with already-existing vampire beliefs in those places, and especially about the intensity of affect and meanings associated with blood. In particular, it can be argued that blood acts like a dream-thought, such that **blood-work** produces a range of surrogate images that can stand for (the taking of) blood: for example, the colour red. Blood-work can also make blood itself stand for specific meanings: white colonial exploitation of African bodies, for example. The colonial authorities worked hard to stamp out these vampire stories, but rarely succeeded. After all, it makes sense that blood taken at the King George V hospital would be taken to England where it would be given to an anaemic-looking King George V.[14]

It would be a grave mistake to assume either that vampire scares are solely based on African superstitions or irrationalities, or that these stories can simply be put down to their premodernness or unmodernness. For vampires come to inhabit modern cities too, including a city renowned for being the capital of Empire, of the Nineteenth Century and of Modernity – Paris. In 1857, the Parisian poet Baudelaire writes of his encounter with a female vampire.[15] Their meeting is erotically charged: she is fragrant, her 'mouth of strawberry', her lips moist. After their passionate ardour was over, after she had 'drained the marrow out of all my bones', Baudelaire found the beautiful vampire had turned to bones (1857a, page 255, lines 21–25):

> Frozen with terror, then, I clenched both of my eyes;
> When I reopened them into the living light
> I saw I was beside no vampire mannequin
> That lived by having sucked the blood out of my skin,
> But bits of skeleton, some rattling remains [...]

It is possible to read this poem as an account of a distressing erotic encounter or perhaps a night spent with a prostitute. Such an interpretation would tally with Benjamin's reading of Baudelaire as a figure engaged with, but alienated from, the modern city.[16] For Benjamin, emblematic of this alienated enchantment with the eroticism and the crowds of modernity is Baudelaire's poem, 'To A Woman Passing By' (1857b). In this short poem Baudelaire describes passing a beautiful women. For an instant, their eyes meet. They both know they could fall in love, but they will never see each other again. In this context, the vampire is an overdetermined metaphor that points in several directions at once: towards erotic encounters with women in the city, towards a fear of feminine sexuality, towards an alienation from eroticism and sexuality in the modern city. This alienation is represented by the moment when Baudelaire opens his fearful eyes and sees that the female vampire is but rattling bones. Baudelaire's description of the vampire before and after her metamorphosis is also significant. She is beautiful, with moist strawberry mouth and a sweet smell. This image of the female vampire may have its origins in Malay culture, as we will see below.

The female vampire can represent the horrors of modern urban sexuality, but rumours about vampires also convey fears about strangers in the modern city. An example of this occurred in London in the late 1960s, where scare stories led to a vampire hunt in Highgate Cemetery (Ellis, 1993). Highgate Cemetery is perhaps best known for its famous dead, among them Karl Marx. However, according to Ellis, the cemetery has also been associated with occult rituals and 'black' magic since at least the 1850s (1993, pages 14–16). In the 1960s, these rituals led to a series of witchcraft scares, especially relating to the abuse of children. Also at this time, Sean Manchester of the British Occult Society was arguing that London's West End, focus of several previous vampire plague scares, was home to a stranger from Wallachia, a famous Vampire (could it be Dracula?).

Gripped by tales of vampires in London, on Friday 13 March 1970, a crowd of vampire-hunters gathered outside Highgate Cemetery (Ellis, 1993, pages 23–26). The police attempted to prevent the crowd from entering the cemetery, but about 100 vampire-hunters succeeded in getting through police lines and began searching the grounds. The vampires, however, were long gone. Although many treated the vampire hunt as a joke, according to Ellis some of the participants were deadly serious. The work of the devil is not to be taken lightly. Later, on 1 August 1970, three girls found the burnt and headless body of a woman outside a vault in the cemetery: it looked like a vampire killing. This was not the only incident that year. To the list of vampire plagues, we should add London, 1970.

Vampires, then, have been around for some time, at least since the time of myths and legends, in which, according to this evidence, moderns still live. In the examples above, it is important to recognise the rationalities behind the figure of the vampire. One common context for these vampire stories seems to be the fragility of the social settings within which they emerge: whether these social settings are in the contact zones between Empires or at the edges of Empire or in places that are in some kind of crisis. Power relations, especially those involving the transmission of knowledge, are clearly important: vampire tales are often a weapon of the weak, acting as a warning, a diagnostic and a call to action. It is possibly for this reason that Karl Marx regularly used the figure of the vampire.[17] As a diagnostic, it shows directly how it is that the powerful live off the bodies of the less powerful. Metaphorically, they suck the living blood out of them, brutally, albeit in shadowy ways. In this sense, the vampire is also a social figure that allows powerful emotions to be vented: for example, by converting a system of generalised exploitation and cruelty into an identifiable and defeatable enemy, albeit one that is illusive and frightening.

The vampire describes a horror-filled world. Their uncanny appearance discloses terrifying relationships that are normally hidden: for some, these hidden relationships stem from childhood anxieties, such as anxieties associated with the mouth and feeding;[18] for others, they can be produced by real traumas in ordinary life, such as disease or fearful strangers or exploitative élites. What is interesting is that vampires are an effective means to simultaneously embody, conceal and allay these traumas and fears, from whatever source. It is perhaps for this reason that vampires have been a popular subject for writers and poets: this list of those attracted to the figure of the vampire is impressive (see Frayling, 1991, pages 42–62), both before and after Bram Stoker's classic tale. And film-makers, of course.[19]

The vampire might have always been with us, but Stoker's tale of Dracula will show us exactly how the vampire finds a home in the horror of modern city life. More than this, we can appreciate how an entire age – an age we sometimes call global and, less often, imperial – might better be thought of as vampiric. Let us turn our attention, now, to Dracula and his geographies.

3.3 London, Dracula and the edges of Empire: circulation, blood, technology

Much has been made of Stoker's background research into the history and geography of Transylvania and of Vlad the Impaler (also known as Vlad Dracula, Voïvode of Wallachia). Bram Stoker's notes show that he used a variety of sources, including: William Wilkinson's *Account of the Principalities of Wallachia and Moldavia* (1820); Charles Boner's *Transylvania: its products and its people* (1865), which contained a pull-out map of Transylvania by E. A. Bielz; Baedeker's *Southern Germany and Austria, including Hungary and Transylvania* (1880); and Emily and Dorothea Gerard's *Transylvanian Superstitions* (1885). *Dracula* is, thereby, set against solid research undertaken in the British Museum and Whitby Public Library. From this evidence, Stoker is able to produce seemingly authentic first-person accounts of the world. Thus, Jonathan Harker, on his journey to Dracula's castle,[20] records the picturesque sights:

> All day long we seemed to dawdle through a country which was full of beauty of every kind. Sometimes we saw little towns or castles on the top of steep hills [...] At every station there were groups of people, sometimes crowds, and in all sorts of attire. Some of them were just like the peasants at home or those I saw coming through France and Germany, with short jackets and round hats and home-made trousers; but others were very picturesque. The women looked pretty, except when you got near them [...] The strangest figures we saw were the Slovaks, who are more barbarian than the rest, with their big cowboy hats, great baggy white trousers, white linen shirts, and enormous heavy leather belts, nearly a foot wide, all studded over with brass nails. (Stoker, 1897, page 11)

Stoker distils and, for his purposes, distorts the best of – what we would now call – Orientalist knowledge of the region and its peoples: a colonial geographical imagination underpins *Dracula*. In many ways, Stoker exoticises and romanticises Transylvania. In so doing, he makes it different enough (from England; from the self-appointed 'civilised' world) to make the existence of Dracula – the vampire – plausible enough, or at least undeniable enough. The dramatisation of the narrative, and also of desire and anxiety, depends upon other geographies too, especially journeys, geographies of communication and sites of danger. To begin with, the book is structured by several long journeys: the journey Jonathan Harker makes to Dracula; Dracula's journey to London; and the pursuit of Dracula as he flees from London to his castle. In these journeys, however, we can see that the 'distance' between Transylvania and England had been collapsed in various ways: legally, financially, physically.

The extension of the railway network across Europe had effectively brought far-off-places (far-off in both space and time) nearer to the capitals of western European Empires, especially Paris and London. Railways also made it possible to travel more quickly and more reliably than before. Similarly, advances in sea travel enabled Dracula to envisage transporting himself – and his precious cargo of coffins, earth and money – safely by ship. This shrinking of the world serves not only to put Empire in closer contact with its peripheries, it also means that the edges of Empire are in closer contact with the beating heart. It is the pulsing blood of Empire that Dracula senses. Dracula's time in feudal society is nearly up: the peasants are on to him and they have a whole variety of folk-weapons to use against the vampire: from Christian symbols (holy water, crucifixes), to homeopathic remedies (garlic) and precautionary measures (not going out at night). Through a variety of means, Dracula learns that his life in London would be so much better. Harker reports Dracula as saying that, from friends,

> I have come to know your great England; and to know her is to love her. I long to go through the crowded streets of your mighty London, to be in the midst of the whirl and rush of humanity, to share its life, its change, its death, and all that makes it what it is. (Stoker, 1897, page 31)

Through Harker's skill in conveyancing, Dracula intends to purchase a variety of properties in London. This empowers Dracula to relocate himself, and also the coffins and earth he needs, to the great metropolis. Only slowly does it dawn on Harker what is actually going on. Only slowly because it is almost impossible for Harker to believe that this strange man, this stranger, is a vampire. In Transylvania, Dracula's strangeness – his undying agelessness – mark him out. He is recognisable and known: this makes it both more and more dangerous and harder and harder for him to continue to live in his castle. In London, it will be entirely different. London's great crowds will act as the perfect cover for Dracula. In a city of strangers, the vampire will be just another man: just another stranger in a great tumult of blood. The city's indifference to him will be his greatest asset and its greatest weakness. Dracula explains:

> Here I am noble; I am *boyar*; the common people know me, and I am master. But a stranger in a strange land, he is no one; men know him not – and to know not is to care not for. I am content if I am like the rest, so that no man stops if he sees me, or pause in his speaking if he hears my words, to say, 'Ha, ha! a stranger!' I have been so long master that I would be master still – or at least that none other should be master of me. (Stoker, 1897, page 31)

Despite his ability to turn himself into other forms, such as a mist or various animals, Dracula is simply too easily recognised in his home-place. He is too easily contained and mastered in the surroundings of his castle. London, on the other hand, offers him new life … or, better, new death. The story, then, involves one of migration – of circulation. Through careful planning and some luck, Dracula succeeds in travelling to London (via Whitby). Though Dracula can move almost invisibly, modernity ensures that he leaves a traceable paper trail. It is this that the vampire-hunters will use to track Dracula down and chase him from London back to Transylvania, where they kill him. The vampire-hunting gang famously include Mina Harker (née Murray), Jonathan Harker, Dr Jack Seward, Quincey Morris, Arthur Holmwood (later Lord Godalming) and Dr Abraham Van Helsing.

For my purposes, there are two aspects to this vampire-hunting worth noting. First is that these characters are continually moving around, so Dracula's is not the only significant journey. In particular, we see Van Helsing moving large distances, from Amsterdam to Whitby to Amsterdam and back again. Second is the paper trail. It reveals the nature of the detective in the nineteenth-century city.[21] On the one hand, the detective has to find the clues.[22] On the other hand, these clues have to be interpreted correctly. In this process, Mina Harker plays a vital role: it is she that gathers and collates the evidence, she who interprets and plans. Without her, there is little doubt that the vampire-hunters – for all their good hearts and hot blood – would probably not succeed.[23] Mina's heroism lies in her unflagging attention to detail. Arguably, this intense attention mirrors Stoker's own.

Stoker's noted obsession with railway timetables and his deliberate use of dated records means that the novel is evidence of precisely the kinds of circulation that were becoming possible in late Victorian England at the height of Empire. This involved, for sure, the phantasmagoric movement of people, goods, ideas, money and information.[24] But Empire also pumped blood around its circulatory system. Clearly, Dracula had already fed on the varied diet that had been available in Transylvania, but now he would be able to feast unnoticed on the international, cosmopolitan fare of London.

On his arrival in England at Whitby Bay, Dracula feeds first on Lucy Westerna, fiancée of Arthur Holmwood and close friend of Mina Harker. He does so by appearing to her at night, perhaps in her dreams. Lucy gradually weakens and sickens. Arthur asks for aid from his friend, Dr Seward. Seward, in turn, calls upon one of the world's most open-minded and advanced scientists, Professor Abraham van Helsing of Amsterdam. Fearing the worst, van Helsing opts initially for a modern medical procedure, the blood transfusion, to overcome ancient evil. Night after night, good English, American and Dutch blood is pumped into poor Lucy's ailing body. Despite the transfusions, Lucy dies – a misfortune that will cause further grief before it is avenged.

It is not just blood that circulates, information does too. The transmission of information is evident in the structure of Stoker's novel: the tale is told through scraps of various kinds of evidence. These are taken from diaries, letters, newspaper cuttings, a ship's log, memoranda, phonograph recordings and telegrams. And, presumably, if there had been the technology to do it, we might even have seen the Kodaks that Jonathan Harker took! Clearly, some of these bits of information travel further than others: telegrams covering great distances quickly and reliably. Presumably, if Stoker were writing today, there would be glimpses of email, mobile phone text messages, faxes and the like. Dracula, moreover, would probably have travelled by plane.[25] The circulation of legal documents, money, and messages shows that Empire was an open and unbounded system, in which a vast array of circuits of people, information, goods, property, money and … blood were operative.

It is also through these fragmentary and transitory transactions that the vampire-hunters are able to track down Dracula. Like modern detectives, they follow the trail: asking workers where various items were taken (e.g. Dracula's coffins); investigating houses Dracula had purchased. In the end, the vampire detectives are able to completely uncover Dracula's vast dark network of properties and coffins, and they are able to save London from the vampire plague about to break out in its crowded and unprotected streets. In this, Dracula is surely a metaphor for all plagues but, as much, he is also a metaphor for intense anxieties about the vast numbers of strangers that were tramping London's streets. This, then, is the lesson of Dracula for understanding the imperial city: the imperial city does not simply belong to the coloniser, in a very real way it also belongs to the colonised and, further, to the foreign world beyond.

Indeed, Dracula's own colonisation might even be said to resemble Bram Stoker's move from the periphery of Empire – Ireland – into the heart of Empire. Bram Stoker was hardly the only, or the first, Irishman to settle in Britain.[26] The imperial city is, in many ways, the archetypal global city: open to the world and to what the world will bring with it. What Dracula shows is that the Empire can bite back – at its very heart. Dracula attempts to colonise the city of the colonisers. The terrible truth that Dracula offers is that he – the denigrated other – is more powerful than the English, than their American cousins and Dutch friends. It is not that Dracula brings with him inferior blood, but that he is better at making use of blood than the English. Dracula explains:

> We Szkelys have a right to be proud, for in our veins flows the blood of many brave races who fought as the lion fights, for lordship. Here, in the whirlpool of European races [...] till the people thought that the were-wolves themselves had come. (Stoker, 1897, page 41)

Dracula is superior not because his blood is purer than that of the English, but because he contains so many brave bloods. In this, he outdoes the English: Dracula's mix is better than theirs, for none is of pure blood.[27] The fear is not so much one of racial mixing, but of not making the best of cosmopolitan blood lines. The explicit anxiety is about blood and race, but implicitly there are anxieties about inter-racial sex and about the deterioration of races in the modern city. This deterioration of race was strongly associated, in nineteenth-century England, with the supposed inferiority of the working class. In cities, this was evident in their (impoverished) living conditions, in their (dirty) bodies, in their blood and death. For *Dracula* is also a diagnosis of the mortal city: foul with blood, feeding off money.

3.4 Mortal Cities: Dracula, blood and money

The nineteenth-century English city was, by most accounts, a pretty foul place. Whether you read Engels or Dickens, the nineteenth-century city is sick, dirty, violent and malodorous – at least, in its poorest parts (see, for example, Engels, 1845; and Dickens, 1848). You can get a sense of this from Engels' description of Manchester as he saw it from Ducie Bridge:

> The view from this bridge, mercifully concealed from mortals of small stature by a parapet as high as a man, is characteristic for the whole district. At the bottom flows, or rather stagnates, the Irk, a narrow, coal-black, foul-smelling stream, full of débris and refuse, which it deposits on the lower right bank [...] Above the bridge

are tanneries, bonemills, and gasworks, from which all drains and refuse find their way into the Irk, which receives further the contents of all the neighbouring sewers and privies. It may be easily imagined, therefore, what sort of residue the stream deposits. Below the bridge you look upon the piles of débris, the refuse, filth, and offal from the courts on the steep left bank [...] Here the background embraces the pauper burial-ground, the station of the Liverpool and Leeds railway, and, in the rear of this, the Workhouse, the 'Poor-Law Bastille' of Manchester, which, like a citadel, looks threateningly down from behind its high walls and parapets on the hilltop, upon the working people's quarter below. (1845, page 89)

As is well known, the poverty, disease and violence of life for the poor provoked a wide variety of responses from the ruling élites. Many of these responses involved sanitary reform, including, for example, the building of modern drainage and sewage systems,[28] not the least part of which included the regulation of filth and odour.[29] Indeed, as Stallybrass and White have argued, the city was treated as if it were a human body (1986, Chapter 3): it was to be cleaned, it was to breath fresh air, and it was to behave itself.[30] Following Alain Corbin,[31] Patrick Joyce has described this as 'civic toilette': 'the cleansing, clearing, paving, draining, and ventilating of the city' (Joyce, 2002, page 106). For Joyce:

The civic toilette was about controlling the markers of organic time so that time might be endured, the markers of death; those of blood, excrement, secretions, rotting and dead matter. It therefore involved the deepest fears and anxieties of those who governed and those who were governed. (2002, page 107)

As Engels' description shows, the city's waterways ran with refuse, human sewage and industrial waste. The slums were crowded, almost uninhabitable, inhuman places, fit only for animals. It was a place of death, marked by blood. Joyce focuses attention on the bloody side of London in the nineteenth century:

This city ran with blood. Smithfield animal market was at its bloody heart: within one year [in the middle of the nineteenth century] well over a million sheep and a quarter of a million other beasts went there on their way to be slaughtered throughout the city [...] The dead were also buried in the centre of the city. The middle of London was likened to a charnel house, the dead piled high upon one another, and constantly present to the living, in the most disturbing forms [...] What happened in the course of the nineteenth century was that death, and the corrupt and decaying bodies of the dead, and this goes for the human as well as the animal dead, were finally rendered invisible and anonymous. (Joyce, 2002, page 107)

An important sanitation strategy was to push those places where the human dead were handled (either by burial or cremation) and where animals were slaughtered and butchered to the edges of the city. This can be seen in Paris as much as London. Blood was purified from the streets of London and other cities. Civic toilette cleansed the body and the city. Controls were instituted over how human and animal dead bodies were to be handled. Clear and precise boundaries were established defining where blood could – and could not – be spilt (e.g. within the licensed slaughterhouse, not on the pavement outside), and these boundaries were policed to ensure against transgression.

In the nineteenth-century city, there is an intense desire to rid the streets of blood, yet, as this is impossible (since the city needs to spill blood to live), particular spaces (such as slaughterhouses) are proscribed where blood can be shed.[32] The vampire transgresses these new orders of blood, both by treating human blood as food and by crossing over the thresholds between the impure and the pure, between the foreign and the familiar. In the figure of Dracula are crystallised anxieties not only over blood but also over race, class and the disgusting city. The imperial city is, then, an anxious place: anxious about what it brings close to its heart, about the human condition of living and dying, but anxious too that its life-blood will be sucked dry, that it will become vulnerable to a world more ruthless, hungrier than itself.

Anxieties about the ruthlessness and hunger of city life were hardly being assuaged by the rise and intensification of the capitalist urban economy. In the late nineteenth century, it was becoming obvious that the simple ownership of money was not enough either to gain power or to secure one's every dream. Instead, money had to be made to work, by putting it into circulation. Jonathan Harker, at one point found 'a great heap of gold in one corner – gold of all kinds' (1897, page 62). The money is useless just lying around, so it is of symbolic importance that Harker employs a gold coin as an escape tool while fleeing from Dracula's castle. Dracula also knew it was no good the cash just sitting there, he intended to invest the money in properties across London. For him, though, the money had no intrinsic value. It means nothing to him to throw gold coins at his pursuers as he makes his own escape from London. Such actions give us a sense of the different values bound up in money. As Simmel argues:

> The notion that the economic significance of money results simply from its value and the frequency of its circulation at any given time overlooks the powerful effects that money produces through the hope and fear, desire and anxiety that are associated with it. It radiates these economically important sentiments, as heaven and hell also radiate them, but as pure ideas. (1900, page 171)

As Dracula circulates, so does his money. However, the value of his gold coins is associated with the hopes and fears, the heaven and hell, that radiate out from them.[33] Just as money radiates economically important sentiments, so does Dracula. The vampire embodies pure economic ideas, such as: the importance of long-term investment strategies; ensuring the security of the money supply; thinking about property far into the future; and the immortality of money-in-circulation in the capitalist economy as opposed to the mortality of money itself. The undead immortal vampire reveals the double-sided character of money: on the one hand, the imperishability of money-in-circulation (i.e., capital) and, on the other, its deathless and life-sucking qualities, such is the eternal cold of gold coins. Marx put it this way:

> Money attempted to posit itself as *imperishable value*, as eternal value, by relating negatively towards circulation, i.e. towards the exchange with real wealth, with transitory commodities [...] Capital posits the permanence of value (to a certain degree) by incarnating itself in fleeting commodities and taking on their form, but at the same time changing them just as constantly; alternates between its eternal form in money and its passing form in commodities; permanence is posited as the only thing it can be, a passing passage – process – life. But capital obtains

this ability only by constantly sucking in living labour as its soul, vampire-like. (Marx, 1857–58, page 646)

Money is vampire-like when it is like Dracula, in undead circulation, constantly sucking the life out of the living. Stoker's *Dracula* places London at the heart of imperial, undead circulations – of people, of money, of anxieties, of blood, of evil. This circulation is immortal in the sense that it keeps coming back: circulating, returning, vanishing, appearing. However, this is never a smooth or continuous or linear process. The vampire – like money, like ideas – can lie still, dead, for long periods of time before becoming unexpectedly active once again. In this sense, the undead or immortal in city life is not necessarily something that is always there, but something that is always threatening to return. The return of the vampire, its inhabitation of the city, invokes the heavens and hells of the city, of blood and of money. In the next section, we will see that these heavens and hells extend to sex and sexuality.

3.5 Singapore, Red Lips and Sharp White Teeth: sex and death

The eroticism of vampire stories exposes some of the forms of desire bound up in the phantasmagorias of city life. Dracula does not simply devour his female victims, he beguiles them. Through hypnotism, the victims lose their ability to fight back. Dracula is irresistible: not through force, but through seduction. His victims eventually give themselves up willingly to Dracula's insatiable desire for living blood. In the moment of seduction, the vampire becomes desirable – the female victims giving him their all. Thus, the vampire becomes a figure of desire: a gendered and sexualised figure, revealing perhaps something of the perversity of sexual desire in city life.[34] The warning 'be careful what you wish for' rang through the last Chapter. In this section we

Figure 3.3
Queen of the Damned, as advertised on Orchard Road, Singapore.

will be worried about *who* you wish for. Here, I will argue that vampires can tell us about *the reversals of desire* in city life. To get at this, I will discuss stories about female vampires in Singapore (Figure 3.3).

One myth in Singapore concerns vampires that commonly go under the Malay name of pontianak.[35] As Carole Faucher (2004) shows, these stories are very popular and take a wide variety of forms. I would like to focus on one particular pontianak story, since this seems to be typical of many versions of the story. It appears in one of the numerous serialised anthologies of horror stories. These series have titles like *There are Ghosts Everywhere in Singapore* and *Ghosts of Singapore*. These anthologies include authored works, but many are anonymous, while some are (seemingly) sent in by readers of such collections. Many of the books are thematised, dealing with everything from soldiers' tales to exorcists, Voodoo to vampires. In most of the anthologies, the stories are short, lasting a few pages or just a few paragraphs.

In fact, what is of interest is the sheer quantity of the tales. The relentless relating of often similar horror stories furnishes them with a kind of force. By sheer weight of numbers, readers are almost cowed into submission – forced to doubt their own doubt in the existence of these horrors. More than this, stories are often told from other places in the world, so we hear of vampires in California, ghosts in London, Voodoo in New Orleans, and so on. Other stories have no distinct location and could be set almost anywhere in the world. The overall effect is to enshroud the reader in the occult, a realm only partially perceptible in ordinary life. Stories are told from a variety of perspectives, some in the voice of the victim, others from the perspective of witnesses – many tell of close shaves with the vampire, and sometimes of the experience of seeing someone close turn into a vampire. With so many different voices, it feels as if there is a crowd of witnesses. Some voices may be better at telling tales than others, but the sheer artlessness of most of the stories adds to their plausibility – they can't all be concocted by crafty (ghost) writers, can they?

Vampires – or, rather, pontianaks – appear in almost every collection.[36] Let us look at a typical tale, the *Lone Woman* (Anon., 2000a). As in many such tales, the story begins with an unnamed man, driving home late at night after work. On his way, he spots someone by the roadside.

> **The woman wore a long red dress and was standing by herself. Had the thought of a lone woman standing by herself at that time of the night not made him think of how dangerous it was for the women in question, he would have driven straight on. But instead, he slowed down and decided to stop for her. (page 30)**

The woman gets into the front seat, without being asked, and somewhat unexpectedly. She smiles at him and off they set. The man begins to wonder why the woman had not been able to hail a taxi, since there were many around. She smiles enigmatically. And, soon, she falls asleep:

> He could not help but look at her because she was so beautiful. She smiled in her sleep. (page 31)

He notices her black hair, her white skin and her red lips.[37] He is beginning to get distracted by his desire for her. Suddenly, he feels her hand on his thigh. Since she's asleep, it must be innocent, so he decides not to move her hand in case he wakes her up. The more the man looks at the women, the more he thinks there's something disconcerting about her. Then, he starts to remember

Stories about young women who were actually monsters, and preyed on men who worked outdoors late at night, were playing back and forth in his mind. They would seduce them and then they would kill the victims to feast on their blood. (page 32)

The man takes a closer look at the woman – using his rear-view mirror – and is alarmed to see that the woman has two razor sharp teeth protruding from her mouth. In his fear, he begins to drive erratically and the woman begins to stir. She smiles, and holds his thigh more firmly and lapses back into sleep. Terrified, the man stops the car and jumps out. Bemused, the woman also gets out and quickly walks to the other side of the road. The man looks around, but she is nowhere to be seen. She has disappeared. Relieved and still afraid, the man drives off and heads back for home. He drives and drives until he uncannily ends up back at the exact same place where he gave a lift to the woman. A shudder of fear runs down the man's back as he realises the pontianak is there, again, waiting to be rescued by some other unsuspecting motorist.

In other pontianak tales, the 'lone woman' is in a white dress, but she is always young and beautiful, and often becomes more and more seductive the longer the man looks at her. Her appearance is usually associated with fragrant smells, especially frangipani. She usually falls asleep, but her dreams are not innocent: in fact, the movement of her body and the sounds she makes are clearly – to the (potential, male) victim – erotic, sexual. To survive, the man has to resist severe temptation. Usually, the pontianak gets put out of the car near a cemetery or beneath an old tree. Frequently, in her irritation, the lone woman throws a large amount of paper money at the driver. Later, he discovers, it is just dried leaves. Occasionally, the vampire shows her true form: turning into an old monstrous woman, who threatens to kill the man and drink his blood.

Figure 3.4 A car park attendant's kiosk, Orchard Road, Singapore.

Vampire women do not just hunt by the roadside, however. They are to be found in apartment blocks, in hospitals, in old places; including cemeteries of course. And in car parks: in one story, an Orchard Road car park attendant, sitting innocently in his kiosk, is attacked (Figure 3.4). The traditional homes where pontianaks 'live' are trees, probably as a result of the root system coming into contact with the (un)dead, especially where the trees are near cemeteries. Often trees disguise the presence of a cemetery; almost too late the victim will

Figure 3.5
A grassed-over cemetery, Singapore.

realise where he is. Sometimes the successful escapee will later be shocked to learn that the grassy patch of ground, where the pontianak had lured him, is actually a disused cemetery (Figure 3.5).

These vampires clearly prey upon men who lay themselves open for (sexual) temptation: give in to her seductions and he is a dead man. In some ways, we can take these tales to be cautionary – and even as a clear set of rules and prohibitions. The man usually knows the stories about dangerous women, but he fails to remember, or act on, these *soon enough* to save himself. Other tales seem to offer advice – what to do if ... you find a loved one craving (your) blood, or if ... you meet a pontianak. What to do? Run, and don't look back!

It is not just in Singapore that vampire women prey on unwary men. In Paris, you'll recall that Baudelaire had his own disturbing encounter. In *Dracula*, meanwhile, Jonathan Harker made this promising, yet unsettling, discovery:

> In the moonlight opposite me were three young women, ladies by their dress and manner. I thought at the time that I must be dreaming when I saw them, for, though the moonlight was behind them, they threw no shadow on the floor. They came close to me and looked at me for some time and then whispered together. Two were dark, and had high aquiline noses, like the Count's [...] The other was fair, as fair as can be [...] All three had brilliant white teeth, that shone like pearls against the ruby voluptuous lips. There was something about them that made me uneasy, some longing and at the same time some deadly fear. I felt in my heart a wicked, burning desire that they would kiss me with those red lips. (Stoker, 1897, page 51)

Fortunately (or unfortunately?) for Harker, Count Dracula intervenes:

> How dare you touch him, any of you? How dare you cast eyes on him when I had forbidden it? Back, I tell you all! This man belongs to me! (page 53)

Indeed, men really do seem to belong to the vampire. Men, in the shadowless shadow of the vampire are constantly on trial, constantly tempted by the seductions of the vampire – and these seductions are almost innumerable: beautiful, erotic, aristocratic, immortal ... need I go on? Men, it seems, have to be ever wary of beautiful young women they meet late at night. The figure of the female vampire can be interpreted as representing the ability of women to castrate men.[38] Such a line of argument is similar to Copjec's argument that the vampire represents repressed infantile anxieties. This time, however, the vampire represents male anxieties about castration (following Freud, 1926). In this view, the popularity of the vampire arises from its expression and transgression of (Victorian and/or Singaporean) sexual taboos.[39] Mighall summarises this position:

> The vampire is monstrous not because it is a supernatural being which threatens to suck the protagonists' blood and damn their souls, but because at some 'deeper level' it symbolizes an erotic threat. (1999, page 211)

The intensity of this erotic threat is evident when it dawns upon Dracula's pursuers that his first victim, sweet Lucy Westerna, might herself now be a vampire. The vampire-hunters resolve to go to Westerna's tomb to examine it and her body. Her body, however, was not there. Seeking definitive proof, Dr Seward and Professor van Helsing return during the day. On opening the coffin, they see a white figure:

> There lay Lucy, seemingly just as we had seen her the night before her funeral. She was, if possible, more radiantly beautiful than ever; and I could not believe that she was dead. The lips were red, nay redder than before; and on the cheeks was a delicate bloom. (Stoker, 1897, page 240)

Van Helsing peels Lucy's red lips back to reveal sharp canine teeth. The vampire-hunters are now convinced that Lucy has herself become a vampire. Now, their duty is to dispose of Lucy, despite the after-life of her beauty. Van Helsing reminds Seward that Lucy is dead and advises him that she has been in a kind of Un-Dead sleepwalk since first being bitten by Dracula. To kill her now would be like killing her in her sleep. Even so, the thought turns Dr Seward's 'blood cold' (page 241). Details of the disposal of the body could hardly have made him feel any better. Lucy's head was to be cut off and the mouth filled with garlic, while a stake was to be pushed through her heart.

Later, Dr Seward, Lord Godalming, Quincey Morris and Professor Abraham van Helsing return to the graveyard, determined to dispose of Un-Dead Lucy Westerna. Their hearts freeze as they see a white figure in the crypt. Dr Seward describes Lucy's appearance:

> The sweetness was turned to adamantine, heartless cruelty, and the purity to voluptuous wantonness. (1897, pages 252–253)

For a moment, the four men are stunned. Eventually, they recover, refusing to be seduced by her voluptuous smile, the diabolical sweetness of her voice. As a gang, they subdue her, then brutally mutilate her body. In this scene, it is easy to see the sadistic fantasies that female desire provoked in these Victorian men: it was better that Lucy should be dead than voluptuous and wanton. Better, indeed, to cut off her head and rape her body with a wooden stake. The vampire does not separate desire and fear, but embodies them as an intense, almost unbearable, ambivalence. As Moretti puts it:

> Dracula, then, liberates and exalts sexual desire. And this desire *attracts* but – at the same time – frightens. Lucy is beautiful, but dangerous. Fear and attraction are one and the same. (1978a, page 99)

This observation is astute, but it is worth pointing out that this particular formation of intense ambivalence does not quite describe the situation in Singapore. In these tales, the seductive vampire women always escape. There is no mutilation of the female body, no final reckoning. She always lives (dies?) to seduce again. If a large part of the vampire is to do with its erotic threat, then it is important to note where and how this threat is allayed and where and how it is not. Victorian men triumphed over the sexual threats of Dracula and Lucy. Singaporean men, on the other hand, must be ever vigilant.

By now, you will have noticed a certain similarity in the eroticism surrounding Dracula's ladies and the vampire women of Singapore and Malaysia. This is surely no accident. I find it entirely believable when Frayling, very much in passing, states that Stoker noted the existence of vampires in Malaysia (1991, page 298). In many ways, this completes a circuit of vampires. Not only have vampires moved from the edges of Empire – in Eastern Europe and the United States – to its beating heart, London and Paris, they have also travelled across Empire – around Africa, from Malaysia to London and back again. These journeys through imperial cities have not simply been about sucking the life-blood out of the living; there have been seductions too. Red, red lips and sharp white teeth: the vampire woman is as seductive as she is deadly. It would take a man of real courage and will-power to resist. In fact, our Singaporean motorist fares somewhat better than our English lawyer, Jonathan Harker, for he would have succumbed had not Dracula claimed him for himself.

These vampire women illuminate a couple more things about desire and fear in the modern city. To begin with, vampire women caution men about the possibility that they may fall prey to female deceptions and desires. The wise will have listened to the stories and will be able to respond to the possibility that there is more to 'it' than first meets the eye: desire can lead somewhere thoroughly terrifying – a reversal into opposites. Tracking the seductive, frightening vampire women also exposes the undead circulations of these desires and fears, from Malaysia to Stoker's London and to Baudelaire's Paris. In the modern city, where strangers, races and sexes find themselves in close proximity, desire and anxiety sit side by side in the driving seat.

Be warned: being modern is not necessarily going to exorcise the vampire. The city can knock down cemeteries and tear up trees, but vampires are harder to extinguish. With their deep connection to the earth, they can learn to make use of the infrastructure that the modern city throws up: its roads, cars, car parks, hospitals, apartment blocks and so on. Fear is never quite as easily erased as a graveyard. Then, again, nor is desire. The vampire reveals the pleasures of the flesh, of the desire to *be with* the beautiful vampire women. What the vampire women show is that vampires are sexually desirable, yet it might even be desirable to be a vampire: beyond death, beyond fear, yet powerful and alluring. It is this that Louis, late resident of New Orleans, will explain for us.

3.6 To be a Vampire, forever: New Orleans, blood and belonging

In 1803, France sold Louisiana to the United States.[40] Even today, Louisiana sits awkwardly in the union of states that makes the American nation. It is tied, by imagination at least, to its French Creole past. Its ambivalent relationship to that past, and to the present, has fostered a sense of the uniqueness and strangeness of the city of New Orleans. And part of this strangeness is New Orleans' long association with vampire myths. The close connection between New Orleans and vampires begs questions about the relationship between blood and belonging – belonging to a race, to a city, to a nation, to modernity itself. These questions are explicitly explored in Anne Rice's vampire novel *Interview with the Vampire* (1976) and again in Neil Jordan's film of her book (1994).

Interview with the Vampire relates the story of the vampire Louis, who is being interviewed in the present day by an anonymous young journalist. The film is clearest on matters of biography and history. Louis begins his sorry tale: he is 24, the year is 1791, and he is very much alive. At this time, Louis is the wealthy master of a plantation outside New Orleans. But after the death of loved ones, he is unable to come to terms with the grief and yearns for an end to his miserable life. Though book and film differ on the exact nature of Louis' death wish, they both allow full expression to his doomed and melancholic heart. While still grieving, Louis is attacked by a vampire, Lestat, who drains Louis' blood almost to the point of death. But he does not die. Perhaps yearning for death, Louis agrees to meet Lestat another time.

Figure 3.6
St Louis cemetery No. 1, New Orleans.

Lestat and Louis, we can surmise, rendezvous at St Louis cemetery No. 1 in New Orleans (Figure 3.6). This cemetery is located just to the north of the French Quarter and, at the end of the eighteenth century, it was deliberately located just outside the city limits, a stone's throw from the *Place des Nègres* (Congo Square).[41] In fact, the cemetery had been relocated from nearby St Peter's Street after a fire that destroyed four-fifths of the city in 1788.[42] Like Bram Stoker, Anne Rice had done her homework. More than this, Anne Rice knows New Orleans, for the cemetery has a visible presence in the urban landscape.[43] This cemetery makes of New Orleans a city of the dead.[44] In Rice's hands, the cemetery becomes a site – and New Orleans a city – of the dead *and* undead.[45] For, it is here that Louis allows Lestat to turn him into a vampire.

Eighteenth-century New Orleans is an ideal hunting ground for vampires such as Lestat: it is simply full of death, casual and unnoticed. Every year, yellow fever would kill its citizens, but every 20 years or so the plague would take several thousand people. In the summer heat, New Orleans was also prone to fires. While Dracula sought refuge in a city of strangers where he was no stranger than anyone else, the vampires in New Orleans hide themselves in the familiarity and ubiquity of death. The villagers found Dracula out because of the bodies he left behind him. No vampire in New Orleans had any such problem: not only were there a lot of bodies, but no-one would pay much attention to how they died. Perfect cover. Even so, New Orleans in the late eighteenth and early nineteenth century was also a bustling cosmopolitan port city, just like Dracula's London. Louis describes it this way:

> There was no city in America like New Orleans. It was filled not only with the French and Spanish of all classes who had formed in part its peculiar aristocracy, but later with immigrants of all kinds, the Irish and the German in particular. Then there were not only the black slaves, yet unhomogenized and fantastical in their different tribal garb and manners, but the great growing class of the free people of color, those marvellous people of our mixed blood and that of the islands, who produced a magnificent and unique caste of craftsmen, artists, poets and renowned feminine beauty. And then there were the Indians, who covered the levee on summer days selling herbs and crafted wares. And drifting through all, through this medley of languages and colors, were the people of the port, the sailors of the ships, who came in great waves to spend their money in the cabarets, to buy for the night the beautiful women both dark and light, to dine on the best of Spanish and French cooking and drink the imported wines of the world. (Rice, 1976, pages 44–45)

Louis continues:

> This was New Orleans, a magical and magnificent place to live. In which a vampire, richly dressed and gracefully walking through the pools of light of one gas lamp after another might attract no more notice in the evening than hundreds of other exotic creatures – if he attracted any at all, if anyone stopped to whisper behind a fan, 'That man ... how pale, how he gleams ... how he moves. It's not natural!' A city in which a vampire might be gone before the words had even passed the lips. (Rice, 1976, pages 45–46)

Among the feverish movement of the city, the vampire could move freely, imperceptibly. Like London, in New Orleans the vampire could hide *in plain night*: visible, present, but unrecognised in the vast circulating crowd. Like Jonathan Harker in Transylvania, Louis is taken with the picturesque idiosyncrasies of New Orleans. Louis is a romantic, marvelling at the exotic beauty of New Orleans. Though a vampire, he remains tied to mortal endeavours and desires – unlike Lestat, his maker, who sees humans only as prey, as food.

For Louis, the Creole city is a kaleidoscope of languages and peoples, of goods and activities, of beautiful buildings, whether in the Garden District or the French Quarter. However, Louis' beloved French city is set to become an American city and, in doing so, lose its Creole flavour. The transition from colony to state is not met with unreserved pleasure by the vampires. In the interview, speaking of the period just after 1812, Louis mournfully intones (in the film version):

> Years flew by like minutes, the city around us grew, sailboats gave way to steamships, disgorging an endless menu of magnificent strangers. A new world had sprung up around us and we were all Americans now.

For Louis, the magnificence of the strangers is to do with their vivacity and variety. This was not, however, to Lestat's taste. (In the film version) Louis recalls a conversation with Lestat, at the quayside, as passengers disembarked from the steamboat *Natchez* (Figure 3.7):

Figure 3.7
The steamboat Natchez, New Orleans.

LESTAT: Filthy mad tide. Lord, what I wouldn't give for a drop of good old-fashioned Creole blood!
LOUIS: Yankees are not to your taste?
LESTAT: Huh! Their democratic flavour doesn't suit my palate, Louis.

Two ideas are caught up in this dialogue, in part both characters revel in the mixed-up meanings of the word 'Creole'. In New Orleans, Creole can simply refer to Louisiana's colonial period or to someone born in that period. In this sense, Creole could be of any blood, or any mix of blood. However, in common contemporary usage, Creole means someone who is of mixed European and African descent. In this sense, Lestat ambiguously lays claim to both pure and mixed blood. By raising the issue of blood in this way, the word 'Creole' actually serves to purify New Orleans' blood of its democracy. Instead, it lays claim to a slave-owning hierarchy where, ironically, slave-owner and slave blood is mixed. What the vampire dramatises in this instance is the vicious bite of slave-owning: it sucks the blood from the bodies of slaves, ruthlessly, with a sickening overtone of pleasure.

In both London and New Orleans, it is blood that is at stake in the vampire tales. There are clearly anxieties about blood: about its transference, about its mixing, about its democracy, about its circulation, about its force of life and death. As articulated by Bram Stoker, Anne Rice and Neil Jordan, however, these anxieties, while about blood, stretch in different directions. Stoker clearly displays and allays a fear that vampire foreigners will feed with impunity upon Londoners. Jordan is concerned to emphasise the exploitative nature of vampires: they feed off the living. Rice's aristocratic rendering of the vampire adds uncannily to the horror, as beauty and cruelty are folded into the same figure.

Rice's contribution, however, is to give voice to the internal world of the vampire: their life without conscience, consciousness without life. Vampires are doomed and, for Rice, it is this that makes them beautiful: part of being doomed, for Rice, is that vampires live beyond their time, across space. Their doom is to lose their connection to their time and space, to be without a space-time of their own. And they are dangerous for exactly the same reason. Intriguingly, in Rice's tale, those vampires that have, or have made, a home in the world are neither doomed nor beautiful – as exemplified by the vampires that Louis encounters in Europe.

Believing (wrongly) that they have killed Lestat, Louis and Claudia – a child vampire[46] – flee to Europe. They begin a quest to find other vampires, who might tell them of their origins. Influenced by popular stories (such as *Dracula*), they search extensively in Eastern Europe. Instead of finding a vampire much like themselves, Louis and Claudia discover just one wolf-like being. Unfortunately, the animal vampire attacks Louis and he is forced to kill it. Disheartened, after years of searching, the vampire pair end up in Paris – capital of so many things. In Paris, Louis finds a degree of peace, as he believes the city to be the mother of New Orleans. Paris, Louis declares,

> had given New Orleans its life, its first populace; and it was what New Orleans for so long tried to be. But New Orleans, though beautiful and desperately alive, was desperately fragile. There was something forever savage there, something that threatened the exotic and sophisticated life both from within and without [...] so that New Orleans seemed at all times like a dream in the imagination of her striving populace, a dream held intact at every second by a tenacious, though unconscious, collective will. (Rice, 1976, pages 219–20)

New Orleans for Louis sounds remarkably like Paris for Walter Benjamin – a dreaming collective: a beautiful, desperate, exotic, sophisticated, fragile phantasmagoria.[47] What Paris offers Louis, though, is a kinship and belonging: a clear link between mother and son, umbilically linked by blood and culture. But Paris is also a stage for decadence and death: blood is well known to have flowed through its streets.[48] In this context, the *Théâtre des Vampires* was flourishing. In this macabre theatre, real vampires pretended to be humans pretending to be vampires, thereby following the golden rule of the age: hide in plain sight.[49] Louis' encounter with these vampires proves yet another bitter disappointment. Instead of finding kinship with vampires seemingly like him, Louis realises that they are decadent, self-consumed, menacing.

Associated with the *Théâtre des Vampires*, though, is Armand, a 400-year-old vampire. Louis is drawn to Armand, fascinated in part by his sheer age, but also by his ancient grace. (In the film) Louis visits Armand, ostensibly to discover more about his vampire bloodline. Instead, Armand tells Louis about what ultimately kills vampires – they lose contact with the spirit of their age:

ARMAND:	Do you know how few vampires have the stamina for immortality – how quickly they bury themselves of their own will? The world changes. Therein lies the irony that finally kills us. I need you to make contact with this age.
LOUIS:	Don't you see? I am not the spirit of my age. I'm at odds with everything. I always have been.
ARMAND:	But Louis that is the very spirit of your age. The heart of it. Your fall from grace has been the fall of a century.
LOUIS:	The vampires in the theatre?
LOUIS:	Decadent! Useless! They can't reflect anything. But you do. You reflect its broken heart.

At this point, in a moment of homoeroticism, Armand strokes Louis' face and says 'You are beautiful'. Louis' broken heart, his alienation, is his beauty. In some ways, the horror of Louis' position is that he has chosen to become – yet, literally, cannot stomach the thought of being – a vampire. (Be careful what you wish for.) He feels damned, soulless. The irony, as Armand points out, is that this is the condition of the age: soulless, damned, beautiful, cruel; alienated from everything. Unlike the other vampires, Louis is in touch both with modernity, an age at odds with itself, and also with New Orleans, a city at odds with itself. Yet, this contact only leaves Louis feeling alienated, both from modernity and from the city. Louis is as contradictory as his time and place.

The Parisian vampires, meanwhile, live beyond time, in a world of their own making. These decadent vampires, for Rice, are grotesque – for they live outside God's law, outside natural law.[50] Yet, this debased vampire world has its own laws. Foremost among these laws, as Louis is about to discover to his cost, is the prohibition against killing other vampires. On hearing Louis and Claudia confess to killing (as they believe) Lestat, the *Théâtre des Vampires* condemn them to death. Armand, however, rescues Louis: together, they flee to the New World. After more than a century, Louis once again crosses paths with Lestat. Lestat is living off rats and other detritus in a house on Prytania Street, in New Orleans' Garden District, appropriately near Lafayette Cemetery No. 1 (Figure 3.8). Contemplating Armand and Lestat, Louis realises that the true terror of being undead is immortality itself – of not belonging: not to a time, not to a place, not even to a race.

Figure 3.8 Lafayette Cemetery No. 1, Garden District, New Orleans.

In vampire tales, blood is intimately linked to race. Race is embodied in the idea of pure and mixed blood, but also through vampire versus human blood,[51] through history and genealogy, and through kinship and belonging. The vampire raises the spectre of race hatred and plays with it. The vampire undermines any sense of racial hierarchy or racial truth: the truth is simple – all humans are food. The vampire's pleasure in drinking blood permits a certain wry humour to enter the field of racial hatred or racial categorisation, for example by treating racial mixing as if it were the equivalent of fish and chips. As with vampires, the city is no place to start arguing about the purity of anything so impure as 'race': blood flows free, mixing as strangers do. Indeed, for both vampires and cities, it may be that the real horror lies in thinking blood does not flow freely.

The vampire appears to be free, free from the constraints of mortality (excepting unfortunate encounters with stakes or sunlight). This freedom comes at a price: the price of belonging to a race, to a place, to a time. The vampire is doomed. Yet, this has also proved to be part of the vampire's charm. The beautiful, graceful, immortal and melancholic image of the vampire has proved very seductive and popular. Indeed, New Orleans itself has taken on something of this image. It turns out, to a degree, that New Orleans and the vampire are inseparably bound to one another.

3.7 New Orleans and Vampire Tours: selling horror and faith in the vampire

Rue Royale (Royal Street), in the heart of the French Quarter, is significant in many tales about New Orleans. It is where devastating fires have started, murders have taken place, ghosts are seen, and also where vampires have 'nested' (Figure 3.9).

Anne Rice's vampires were not the first – nor even the last – to find a place in New Orleans' immoral, occult atmosphere. In her vampire novel, *Lost Souls* (1992), Poppy Brite describes it this way:

> There had always been New Orleans. Christian had lived in other places, far away across sunless seas, places older and darker and just as strange, with ghosts aplenty.

Figure 3.9 Royal Street, New Orleans. This building was chosen as the location for Lestat, Louis and Claudia's home in the film *Interview with the Vampire*.

> But where else did slave spirits still lament in the Royal Street house of sadistic Madame LaLaurie, where else could one still smell the lingering sweat of a slave woman chained to a stove all the years of her life? Where else did crows flap over the crumbling ruins of St. Louis Cemetery and settle, inky and baleful of eye, on a tomb slashed with hundreds of red *X*'s – *X*'s in faded crimson chalk, *X*'s still fresh and glistening, *X*'s for Voodoo curses, *X*'s to invoke the wrath of Marie Laveau, the Voodoo queen who stayed young forever? (Brite, 1992, page 63)

For Ghost, a character in the novel,

> At first it seemed that there was too much magic here, that it could only cloud intuition and distract faith. On every street corner was another story, in the elegant shade of each courtyard another hovering spirit. Some of them were greedy and reached out to his sensitive mind, whispering *come in, come into*

the vampiric city 123

me, listen to my tale. The buildings and sidewalks themselves seemed to have a susurrant, subliminal voice. (1992, page 307)

Figure 3.10 Vampires at Halloween, New Orleans.

Poppy Brite's vampires – though they are, like Anne Rice's, at odds with almost everything, including mortal moralities – are most comfortable in New Orleans. And New Orleans is happy to have them. Ghost might have been barely able to discern the subliminal voices of the ghosts present in the buildings and sidewalks, but the many (in/famous) vampire tours do much to make their stories absolutely unmissable. By day and by night, throughout the year, New Orleans tells its own vampire tales (see Figure 2.11). In this section, I would like to take a little time to talk about how New Orleans has taken vampires to its heart. Vampires – and, indeed, Voodoo and ghosts – are well and truly part of the phantasmagorias of New Orleans (see Figure 3.10). A quick trip through the French Quarter will leave you in no doubt about this (Figure 3.11).

Figure 3.11 Boutique du Vampyre, New Orleans.

There are a wide variety of walking tours of New Orleans. Poppy Brite observes that each street-corner has a story to tell. Well, it is also true that each street-corner in the French Quarter and Garden District may also have a walking tour party of 20 to

Figure 3.12 Vlad Tepes Knight on a *Haunted History* tour, September 2000.

50 people listening intently to a storyteller – each having paid something like $15 for the privilege. The walking tours are a striking phenomenon of New Orleans tourism. The most popular tours involve guides leading groups around showing them the architecture, history and mythology of New Orleans. Key sites include the large mansions of the Garden District, St Louis cemetery No. 1 and Jackson Square. The Voodoo, vampire and ghost tours are particularly impressive. In fact, one of the entertainments on Royal Street at night can be just watching the different groups – I once counted seven within a one-block stretch – perform a convoluted choreography, as their guides negotiate arrivals, talks and departures.[52] One tour group is called *Haunted History*, run by Sidney Smith. A few years ago, they could boast a real-life vampire, appropriately named Vlad Tepes Knight, as one of their attractions (Figure 3.12).

Rather than give a detailed account of the many vampire sites these tours dwell upon, I would like to focus on just two: first, LaLaurie House and, second, the Ursuline Convent. These sites will give us a sense of the vampiric histories of New Orleans, a context into which Anne Rice's and Poppy Brite's novels are woven. It is worth saying that on the tours key sites in Anne Rice's novels and films are usually pointed out: the *blood-work* that the guides do is to cross-reference the literature, mythology and 'fact' of vampires, creating a dream-like weave for New Orleans in which almost anything, everything, is imaginable.

Figure 3.13 LaLaurie House, 1140 Royal Street.

the vampiric city

125

The best example of the historical 'fact' of vampirism in New Orleans is LaLaurie House, 1140 Royal Street (Figure 3.13). In 1831, the house was inherited by Delphine Macarty LaLaurie, daughter of Chevalier Louis de Macarty, owner of a nearby plantation, and her third husband, physician Dr Nicholas LaLaurie (of Paris). Madame LaLaurie's cruelty towards servants was already well known. She had once chased a 12-year-old child servant, Leah, to her death. The only punishment she received was a $300 fine. However, it is events reported in the local newspaper, *The Bee*, on which most stories about the house rely. On the night of 10 April 1834, there was a fire in the house and the fire brigade rushed to put it out. The fire-fighters could hear screams coming from within the house and had to break through a locked door to get to where they were coming from. What they discovered sent shivers down their spines, a few vomited with the stench of death. Let the *Haunted History* tour's own book take up the story:

> Numerous people were chained to the walls, maimed and disfigured, obvious victims of cruel medical experimentations. Many were dead but some were still alive. Several seemed as if their faces had been disfigured, making [them look] more like gargoyles than humans. One man looked as if he had been the victim of some crude sex change operation [...] Another victim obviously had her arms amputated and her skin peeled off in a circular pattern, making her look like a human caterpillar. Yet another, had been locked in a cage that the newspaper described as barely large enough to accommodate a medium-sized dog. Breaking the cage open, the rescuers found that the LaLaurie's had broken all of her joints resetting them at odd angles so she resembled a crab. (Smith, 1998, pages 18–19)

This version resembles, but does not exactly repeat, the version in *The Bee*. On 11 April 1834, *The Bee* reported that seven slaves were rescued from the building, all of whom had been 'more or less horribly mutilated' over a period of several months. The next day, an angry mob of 2,000 people searched for the LaLauries but they had already escaped, possibly back to Paris, or possibly inland to St Francisville. Since then, periodically (at the time, then again in the 1890s, 1930s and 1970s), a small child (Leah?) in chains, covered in blood, has been seen on the balcony, sometimes running, often falling to the street below. However, other stories of strange equipment found in another room led to speculation about Madame LaLaurie being a latter-day Countess Báthory. Was she drinking blood? Was she bathing in blood? Was there some connection between Madame LaLaurie and Countess Báthory? Indeed, were they the same person?

What interests me is the way these stories are told. Although the tales differ as to the exact nature of the maltreatment of the 'servants' (they are rarely described as 'slaves', let alone 'African slaves'), there is a relish in the way that the details are related. In some ways, this is straightforward theatre – a theatre of vampires. In the stories, there is no sympathy with the victims: indeed, the more victims, the more terrible the injuries, the more cruel the practices, the more blood, the better. The story of the Ursuline Convent is subtly different.

The convent is one of the oldest buildings in New Orleans. In 1725, one of the city's founders, Bienville, invited the Ursuline order of nuns to New Orleans to run a military hospital. Ten nuns, led by Mother Superior M. Tranchepain, arrived in 1727 and established an orphanage and the first school for girls in North America (this school still exists today). The Ursulines were also heavily involved in caring for the sick at times of plague and fever. The story goes that women, in this period, were encouraged to travel to the new colony in New

Figure 3.14 The Ursuline Convent, New Orleans. Are there vampire women behind the shutters on the upper floor?

Orleans. However, instead of paying for their passage, the only requirement was that they bring a coffin with them: most were expected to die quickly, for one reason and another, and coffins were expensive and in short supply in the colony. The sight of women disembarking with only their coffins fuelled rumours, so the story goes, that they were vampires. These newly arrived women were housed first of all in the convent buildings. Onlookers began to notice that none of the windows in the upper floor of the convent was ever opened. Speculation rose that vampire women were being kept in these forever-dark floors. Speculation is that these women are still there, sealed and concealed on the upper floor (Figure 3.14).

There is, in this tale, more sympathy for the women who made the perilous voyage to New Orleans, which in the eighteenth century was just as perilous to inhabit. These tales of easy death, of travelling with your own coffin, of dangerous women forever locked in attics, lend themselves to vampiric imaginations. The phantasmatic association between these vulnerable women and vampires is built on death, on loss of life and, as Copjec might have it, on the real possibility of their deathly desiccation. Once again, as in pontianak stories, fears about female bodies are simply expressed through the figure of the female vampire. This time, however, they are locked (safely?) away in a convent, a closed community of mysterious women.

What is important is that New Orleans thrives on such stories of blood-loss and death. People are eager to hear of the brutalities and sadistic practices of the past. Perhaps the fears represented here are easily allayed in contemporary America: land of the dream. On the other hand, these tales force New Orleans to live with the vampire, to live with the desire for the vampire. Indeed, it is this that the vampire tours have to teach other cities: vampires are here to stay. More than this, vampires are hard to hunt down, they have supernatural speed and find it easy to outlast mortal kind. This leads to a final point, on which I will conclude this chapter – the vampire exposes something of what is immortal about city life and of the ways that cities suck the life out of their inhabitants.

3.8 Conclusion: blood, life and the undead circulations of city life

In some ways, I have been doing the blood-work of cities, to diagnose aspects of their emotional lives. To begin with, we noted the intense ambivalence surrounding blood itself. The form of this ambivalence is far from being transhistorical or transgeographical. New conditions create new forms of ambivalence: a good example would be the blood-work being done by, and on, HIV/AIDS and how it has reconfigured how contagion, sex and death are understood.[53] Condensed in, and displaced on to, the figure of the vampire are desires and fears about sex and death, blood and circulation, purity and danger. Vampires have been intensely ambiguous figures, constantly facing away – as the undead are wont to do – from anything capable of staking them down. Moreover, the very mistiness of vampires has allowed other networks of affect, meaning and power to be traced out: those of kinship, mobility, fluidity, bloodlines, mortality and so on.

Figure 3.15 Vampire wine on Decatur Street, New Orleans.

Vampires are significant figures in the phantasmagorias of city life: be they of blood or of the undead, or even of dreams and magic. As a figure evident in many cities, the vampire can give us a strong sense of the relationship between sex, seduction, mortality, death and even pleasure in any particular city (Figure 3.15). The anonymous pontianak, for example, is a very different social figure from, say, Madame LaLaurie. Similarly, though *Buffy the Vampire Slayer* might indicate a desire (in America) for there always to be a saviour or express the suspicion (of Americans) that suburbia has been built over a hell-mouth, neither of these ideas would necessarily have currency in other places, at other times.

From the edges of Empire, the vampire is now living among the crowded streets of the burgeoning metropolis, where people's movements, and their deaths, cannot be completely accounted for. As with Voodoo across the Atlantic, tracking the vampire between London, Singapore and New Orleans has revealed the specific paths through which things arrive in, inhabit, and leave, the city. I argued, in Chapter 2, that there were occult globalisations working through cities in various ways, actively making them in different ways. Here, I have been concerned with *undead globalisations*: the way that things that do not have a human sense of death – and life – might nonetheless constitute, and circulate through, people's lives. Following the vampire shows that the linkages between cities in one place (say, the west) and another (say, South East Asia) are not always in broad daylight, nor do they exist in mortal times. Vampires jump times and spaces; indeed, they make vampiric times and spaces for themselves. Cities are wide open to these vampire colonisations and globalisations. Cities are vulnerable, we can now observe, to colonisation and globalisation by (undead) forces that seem to exist in worlds far beyond them.

Vampires *reflect*, at one and the same time, the anxieties and desires of the city. We have heard of the modern vampire's insatiable thirst for blood. Vampires have also been sexual creatures: hypnotic and beautiful, graceful and seductive. Under the influence of great cities, the vampire has become urbane, sophisticated, cultured, excited by contact with strangers, revelling in their decadent pleasures, and capable of great sadness. The vampire has made the modern city its home. But is the vampire at home in the modern city? It seems so. Cities bring strangers into close proximity and mix them up. Cities are full of strangers, familiar and unfamiliar, unknown and unremembered. Cities are the natural breeding ground of vampires – eager to suck the life out of their diverse inhabitants, for there is a ready supply of potential victims who will not be missed when they are gone. Perhaps cities have something to teach vampires. Under the influence of cities, maybe the vampire can learn to celebrate human diversity in terms other than blood – maybe then the vampire will appreciate the lives of others and even take responsibility for their blood-sucking ways. This may not be the vampire's choice: it will be up to cities to identify the vampire and to take the consequences of its actions seriously.

That vampires always threaten to evade mortal taming has something to do with their immortality, or rather with the fact that they are Un-Dead. The vampire demonstrates how it is that an immortal perspective can lead to a callous indifference to life, to treating people as if they were animals (or worse than animals). In certain rhetorics, the vampire is analogous to the life-sucking ways of capitalists and of the money economy. Viewed like this, the city itself becomes vampiric: living off the dead labour (as Marx might have it) of its denizens, denizens who are only dimly aware of its vampiric temporalities – even as they themselves are becoming creatures of the night, in the 24/7 city. Indeed, the vampire calculates advantage and profit differently from mortals. As it is with vampires, so it might be with cities. They are indifferent to wealth as mortals know it. Mortals may attempt to accumulate vast wealth in their lifetimes, but vampires and cities can wait, letting moments of crisis pass them by. The danger arises for vampires and cities when they lose contact with their age, with life itself – when they cease to have the stamina to change or adapt.

The Un-Dead Vampire stands between life and death. As a story about sex and beauty, life and death, the vampire has the privilege of being in direct contact with the life of the city, even as it holds the living by the throat. The vampire is a fully embodied social figure, albeit with fangs, hypnotic eyes and red, red lips. Though dead, the vampire has a corporeal life, albeit governed by a cold, ruthless, blood-sucking, soul-less heart. Though dead, the ghost is quite a different social figure. Yes, like the vampire, the ghost stands between life and death. But the ghost is in full contact with neither world. The ghost is intangible, bodiless, governed by seemingly indecipherable imperatives. Unlike vampires, ghosts care about the living, just *not* in ways that the living always appreciate. Unlike vampires, ghosts are reluctant, unable even, to up stakes and move on. Ghosts haunt.

Like magic, ghosts may at first sight appear to have nothing to do with the modern city, but they often appear and reappear in the phantasmagorias of city life – the ghost-like figures that float through the city. Though ghosts usually find it hard to communicate with the living, they are informative. They can teach us about what it means to haunt a city, about the procession of injustices that pass through city life, and about how hard it is to move on. So, it is to the haunted city that we finally turn.

4

The Ghostly City
in which the city is haunted and haunting

> The tradition of the dead generations weighs like a nightmare on the minds of the living. (Karl Marx, 1852, page 146)

4.1 Introduction: the haunting quality of city life

> The ghost is not simply a dead or a missing person, but a social figure, and investigating it can lead to that dense site where history and subjectivity make social life. The ghost or the apparition is one form by which something lost, or barely visible, or seemingly not there to our supposedly well-trained eyes, makes itself known or apparent to us, in its own way, of course. The way of the ghost is haunting, and haunting is a very particular way of knowing what has happened or is happening. Being haunted draws us affectively, sometimes against our will and always a bit magically, into the structure of feeling of a reality we come to experience, not as cold knowledge, but as a transformative recognition. (Avery Gordon, 1997, page 8)

According to sociologist Avery Gordon, the ghost is a social figure that speaks to us of loss, of trauma, of an injustice. What ghost stories can do, moreover, is introduce a haunting affect that permits an emotional recognition of that loss, trauma and injustice. Ghost stories can creep you out, chill you to the bone. They can be unsettling, uncanny, frightening. This is certainly one reason to track the social figure of the ghost through the phantasmagorias of city life. Another reason concerns the ghost's relationship to time and space. By haunting, ghosts betray the significance of time and memory in the production of urban space. Haunting is closely associated with place; ghosts haunt places, spaces or locations. Ghosts rarely move far from the places associated with death, with the site of loss, trauma and injustice. Though closely associated with places, ghosts are not always present.

Ghosts appear and disappear. In doing so, ghosts destabilise the flow of time of a place. They change a place's relationship to the passage of time – as ghosts seem to pop out of nowhere, but are somehow always there. On meeting his father's ghost yet again, Hamlet (in Shakespeare's play) describes this unsettling experience this way: 'The time is out of joint'

(Scene V, Act I).[1] He could, perhaps, equally have said 'This space is out of joint'. Along with a sense of loss, trauma and injustice, ghosts register the dislocation of urban times and spaces.

For Gordon, ghosts impart a creepy, disturbing recognition of the losses, traumas and injustices of city life. The emotional quality of ghost stories can lead to 'transformative recognition', she suggests. Ghosts can be revelatory, progressive. This is not Marx's sense of ghosts. According to him, ghosts tie the living to the traditions of the past and are profoundly reactionary. Ghosts cling to the minds of the living, like a nightmare that hangs around after waking. This is the political question that lies at the heart of this chapter – are ghosts transformative or reactionary? What, then, are we to do about ghosts?

To begin with, I will tell some stories about ghosts in Singapore. The purpose is to set out the haunting relationship between a city's pasts and the location of trauma.[2] The first set of stories relate directly to a traumatic period in Singapore's history – the occupation of the island by the Japanese during the Second World War. Though exhausted, and outnumbered by the defending British and Australian troops, Japanese troops eventually captured the strategic seaport of Singapore on 15 February 1942. So began three and a half years of brutal rule. The ghosts of these actions remain today.[3]

On the night of 14 February 1942, Japanese soldiers stormed Alexandra Hospital, then overflowing with casualties, and began bayoneting patients, doctors and nurses,[4] including those in Operating Theatre No. 1 where an operation was in progress. Those that survived were herded into small, airless rooms. In the suffocating heat, and without water and food, many died. Throughout the night, survivors were dragged out and killed in the hospital grounds. The ghosts of the English victims are often seen, especially in the operating theatre (Figure 4.1). Lt General Yamashita, the Japanese commander, was horrified when he heard of the atrocity and promptly executed some of the soldiers involved. Their ghosts are said to haunt a nearby hilltop.

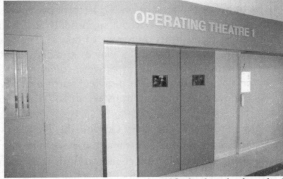

Figure 4.1 Operating Theatre No. 1, Alexandra Hospital, Singapore.

Within Singapore, the Japanese treated the Chinese with suspicion and sought to 'purify' the community of anti-Japanese sentiments. All Chinese males were forced to report to the Japanese military for 'screening', known as the Sook Ching ('purge through purification'). The Japanese military police, or *Kempeitai*, took around about 25,000 young Chinese men to various beaches, including Punggol and Changi, as well as locations inland, and executed them.[5] Today, young Chinese go down to Siloso Beach to

Figure 4.2
Siloso Beach, Singapore.

spend the night, hoping to experience the ghosts that dwell on the beach (Figure 4.2). They report seeing Japanese soldiers coming at them with bayonets, seeing people massacred, and even experiencing the massacre first-hand.[6] As Ann Hui's movie *Visible Secrets* (2001) shows, such visits by young people to haunted beaches also happen in Hong Kong.

The ghosts of 1942 alert us to the relationship between terrible events and specific places within colonial and occupied Singapore. Nevertheless, there are more ghosts in Singapore than this.[7] Indeed, there are ghosts everywhere in Singapore. This is especially so in the seventh lunar month, when, according to Taoist traditions, hungry ghosts are in full contact with the world of the living.[8] At this time, known as the Festival of the Hungry Ghosts (*Zhong Yuan Festival*), people leave out food and drink for the hungry ghosts, usually including something sweet. They also burn joss sticks, paper that symbolises gold, silver and clothes, and specially printed hell bank notes of various denominations (see Figure 4.3).

Figure 4.3
Chinatown during the Festival of the Hungry Ghosts, Singapore.

I asked one man about the hungry ghosts. He told me that he didn't believe in them, as he put more clothes (green joss-paper) on a small, street-side fire. But, he added, some people die in unfortunate circumstances and need food, clothes and money to sustain them in hell. He felt he would owe someone something if he didn't do something for the hungry ghosts every year. Again, he told me that he didn't believe in the hungry ghosts. Then, he told me about a friend of his who had mocked the hungry ghosts one year. The man stood up and bent double, holding his back as if in pain. This is what happened to his friend, he explained. You can't be too sure.

If ghosts are not treated with respect, they are clearly capable of playing tricks on the living, even harming them. In many ways, such an idea accords with Freud's description of beliefs in ghosts. Freud's argument goes like this. The living are fearful of the dead because some part of them is pleased, or relieved, that they are gone. These feelings of pleasure in someone's death cause the living to feel guilt and shame. For Freud, the living *displace* their guilt and shame, and their expectation and fear of punishment, on to the figure of the ghost.[9] Thus, ghosts intend the living harm because the living secretly harbour ill-feelings towards the dead. To appease ghosts and to dissuade them from doing harm, the living must show that they are unequivocally sorrowful about the death and/or are overwhelmingly grateful for the things the dearly departed have bequeathed them.

There is more to the Festival of the Hungry Ghosts, however, than a Freudian fear that ghosts have returned seeking damages. For example, it is believed that ancestors, as ghosts, can influence events in the world of the living. Giving food, money and clothes to the hungry ghosts is an 'investment': the more you give, the more luck you will receive in return.[10] As a result,

Figure 4.4
A small group gather to burn offerings to the hungry ghosts, Orchard Road, Singapore.

many people burn a large amount of joss-paper in their offerings (Figure 4.4). However, the care taken over even the most modest street shrines indicates an honouring of the dead (Figure 4.5), as do the puppet shows and theatre put on to entertain the ghosts while they are visiting the realm of the living (Figure 4.6). Alongside a precautionary attitude towards the dead, there is a deep sympathy with their plight and a generosity of spirit and charity in these acts.

Such beliefs, I would argue, do not simply co-exist with modernity, nor are they some kind of vestigial premodern superstition that will somehow disappear in the modern city; they are part of what it means for

Figure 4.5 Prayers for the dead, Singapore.

Figure 4.6 Theatre for the dead, Singapore.

Singapore to be modern. In other words, Singapore's modernity and its urbanism is ghostly, haunted, just as it is magical and vampiric. Singapore is not alone in this – such experiences with ghosts expose the ghostly figures that parade through the phantasmagorias of modern city life, the hauntedness of modernity itself. Thus, Michael Bell has argued that:

> We moderns, despite our mechanistic and rationalistic ethos, live in landscapes filled with ghosts. The scenes we pass through every day are inhabited, possessed, by spirits we cannot see but whose presence we nevertheless experience. (1997, page 813)

Haunting ought to be antithetical to modernity, yet ghosts seem to claw at the heels of the living, of modernity. Far from ghosts representing a drag on progress, I would argue that haunting lies at the heart and soul of modernity.[11] And the ghost is a familiar, if disturbing, figure in the phantasmagorias of city life. Already, certain ghostly themes have begun to emerge that will add to our understanding of these phantasmagorias. These themes have to do with loss, trauma and injustice; with memory and place; with the dislocations of time and space that ghosts create and represent; with the demands that the dead make upon the living; and with the question of what makes an appropriate response by the living towards the dead.

I will begin with the ways that childhood experiences and involuntary memories come to haunt places. This shows that the procession of ghosts through city life has as much to do with personal experiences, as social histories, as the physical and social life of cities. In **section 4.2**, I will describe Benjamin's account of his childhood in Berlin. This is neither simple autobiography nor a description of significant places in his childhood. Benjamin makes Berlin into a phantasmagoria in which images from his memories, from the city scenes around him, from his dreams, flash up before the eyes. The effect is unsettling, as the ghosts of the past return to shock the living.

The experience of involuntary memory can be profoundly disturbing and frightening. In **section 4.3**, I look at the affect – the uncanniness – of the unexpected appearance of the past in the present as conveyed in the film *The Sixth Sense* (1999). Here, childhood experiences and fears are embodied in the figure of a very disturbed little boy, who may or may not be able to see dead people. Certain locations in cities become known for being haunted. These locations are often associated with injustices in the past, and ghosts appear to represent these injustices. **Section 4.4** begins with a New Orleans ghost story that reveals the brutality and callousness of the past – a past that reappears in the present through its location in an old building. But newer, modern, buildings in cities can also be haunted, as ghost stories set in London and Singapore show. These ghosts, moreover, bring to light the heterogeneity of temporalities in modern city life, the phantasmagorias of time that course through city spaces.

Through their association with specific locations (such as buildings), ghosts access the present – to make their (often incomprehensible) demands upon the living. The living can respond in a variety of ways. In **section 4.5**, I explore the emotional burden that ghosts place upon the living, as they haunt the very streets of the city. Through the emotional chords they strike, ghosts can also haunt the politics of the city. Political questions are explored in the next two sections. **Section 4.6** does so by examining a curious incident: an attack upon a seemingly anonymous, though eye-catching, London office-block by unidentified assailants. While, in **section 4.7**, I show how the dead weigh upon the politics of the living like a nightmare. In the aftermath of an anti-capitalist protest in London, ghosts once again went over the top. The effect was to alter the terrain of political debate from questions about a better future to those about a lack of respect for the sacrifices of previous generations. In both sections, I show that the future is chained securely to a politics of ghosts. This raises vital questions about what is to be done about the city's many ghosts: that is, about how cities can properly honour the dead, without ending up in perpetual servitude to the dead. This is the political dilemma that runs throughout the chapter.

Let us look first at Benjamin's memories of his Berlin childhood, to explore how cities act as sites of involuntary, disturbing memory.

4.2 Remembrance of Times Past: memories, phantasms and Benjamin's Berlin

Memories are a persistent feature of people's experiences of the city and city life. According to Benjamin, memories can flash up in front of your eyes as you walk through the city. Ghosts accompany the living, even while the living are not always aware of their presence. Then suddenly, the past can make its presence felt – the ghost appears. These memories can be drawn from many sources, much like the material drawn upon in dream-work. They can be recent or fantastic, half forgotten or even long forgotten. In this context, childhood memories can be especially intense, especially disturbing, especially haunting. The buried fears and desires of childhood can suddenly be remembered, as the city acts as a trigger sparking a flash-back.

The experience is phantasmagoric, as a motley crowd of memories suddenly surges past – just like a dream. The best place to see this is in Walter Benjamin's description of his Berlin childhood. In 'A Berlin Chronicle' (1932a), Benjamin begins to reconstruct Berlin through a series of reminiscences, about places, about people, about the times and spaces of Berlin. Instead of reconstructing a 'true' account either of the city or of his own autobiography, Benjamin's chronicle weaves in and out of the places of memory.[12] The result is a labyrinthine piece that journeys in time and space, through simultaneously psychological, physical and urban scenes. There is a haunting quality to Benjamin's writing and this is no accident: he is trying to evoke a quality of modern city life.[13]

To begin with, Benjamin starts his tales of Berlin by inviting the reader to meet characters from the past who have held meaning for him. Indeed, he envisages a map of Berlin in which significant relationships are traced out. Moreover, Benjamin also outlines a method of mapping the city – by walking the streets, by losing one's self in the city *as if* losing one's self in a forest (page 298). Giving yourself up to the city, he believed, was about more than immersing yourself in the city, it was also about being able to pay attention to the fragments of city life.[14] From each fragment, Benjamin traces out an association. As with dream-analysis, each association leads to further thoughts, other fragments.[15] In some ways, however, Benjamin's method is as much like archaeology as it is dream-interpretation. Thus, he recommended careful digging through the layers of dust that cover the fragment. Disinterring the dead city, Benjamin argued, would often require returning to a site over and over again. On returning to Berlin, he discovered just how ghostly it was:

> Noisy, matter-of-fact Berlin, the city of work and the metropolis of business, nevertheless has more, rather than less, than some others, of those places and moments when it bears witness to the dead, shows itself full of the dead; and the obscure awareness of these moments, these places, perhaps more than anything else, confers on childhood memories a quality that makes them at once as evanescent and as alluringly tormenting as half-forgotten dreams. (1932a, page 316)

Thinking about his youth, Benjamin recalls his friendship with Fritz Heinle (pages 306–309). According to Benjamin, Heinle was a poet, but he had died aged 19. Benjamin's sadness at the loss of his friend is clear: he remembers his life and passion, his revolutionary fervour. It is not clear how it is that Jewish Heinle came to die at the same time as his Christian girlfriend. However, their taboo relationship in life became intolerable in death: the relatives were unable to find a cemetery that would take the couple. They were segregated in death: matter-of-fact Berlin bore witness to its cruelty in life in the treatment of its dead.[16]

As we saw in Chapter 1, Benjamin took dreams to be a real part of life and evidence of hidden desires and fears. So it is no surprise that his remembrance of Berlin would include dreams.[17] One particular memory is associated with a house, 10 or 12 Blumeshof. Many happy hours of his childhood, says Benjamin, were spent in this house. He was allowed to listen to the piano, read books. Perhaps surprisingly, then, Benjamin tells of an uncanny feeling associated with the building:

> I am met on its threshold by a nightmare. My waking existence has preserved no image of the staircase. But in my memory it remains today the scene of a haunting dream that I once had in just those happy days. In this dream the stairway seemed under the power of a ghost that awaited me as I mounted, though without barring my way, making its presence felt when I had only a few more stairs to climb. On these last stairs it held me spell bound. (page 329)

From this recollection, Benjamin goes on to describe the spatial layout of the rooms in the house. He remembers his grandmother and how he had to cross a large room to reach her. The memories seem disconnected: ghosts, grandmothers, staircases, rooms, going up, crossing distance. The house is a place of memories out of joint, of ghosts of the past, of his way being blocked.[18] Further, the house is represented as dream-like. More than this, Benjamin is using the double insight that significant memories persist in dreams and that significant dream elements persist on waking. However, this particular dream is a nightmare, and the sense that dreams contain, or reveal, some catastrophic memory haunts Benjamin:

> The dread of doors that won't close is something everyone knows from dreams. Stated more precisely, there are doors that appear closed without being so. It was with heightened senses that I learned of this phenomenon in a dream in which, while I was in the company of a friend, a ghost appeared to me in the window of the ground floor of a house to our right. And as we walked on, the ghost accompanied us from inside all the houses. It passed through all the walls and always remained at the same height with us. I saw this, though I was blind. The path we travel through the arcades is fundamentally just such a ghost walk, on which doors give way and walls yield. (1927–40 [L2,7], page 409)

People do not just sleepwalk through the modern city, they ghost-walk through it also. In this light, cities are haunted as much by flesh and blood people as by ghosts. From the dream, Benjamin learns about the production of the arcades as a particular kind of space, haunted by people who are free to move as if there were no doors and no walls. The ghosts allows Benjamin to see the free-floating, uninhibited, phantasmatic movement of people through modern city spaces. At this point, we might begin to think about how contemporary shopping malls create a similarly deterritorialised space, permitting easy movement – once you're inside, that is. The arcade, in this light, creates an interior space that is also an exterior space. The appearance of the ghost is a marker of a threshold: between one world and another. Indeed, the ghost crosses over between worlds, existing fully in neither, unable to find a proper place. Although the ghost appears to be free, it marks a space of dislocation – a space where worlds are inverted, where thresholds can be crossed.

The ghost stands at the threshold of the personal and the social, trafficking feelings and memories across a never clear, and never simply open or closed, border. In the dream, Benjamin recalls that he knew the ghost was there, yet he could not see it. Interestingly, Benjamin does not say that he is terrified, or even scared, by the ghost. Somehow, the affect of terror has been displaced in dream, away from the figure of the ghost. The terrible realisation of the dream lies elsewhere. As Benjamin suggests, it is the people who live in – haunt – cities that are the ghosts. It is as if the living are already dead. And the corollary of this might be that the terror of this realisation is *displaced on to* the figure of the ghost, such that the living can confirm the fact that they are still alive because they know they are not ghosts. They're not, are they? Perish the thought.

Benjamin's ghost floats unimpeded though space, but equally ghosts float freely across time. Even so, ghosts are not caught up in the smooth flow of history, but have one foot in a particular moment or age. They cut across history, jump times, exist in a history of their own. Ghosts are figures that disrupt the linear procession of time that leads from the past, through the present, to the future. Benjamin might have wondered, therefore, both what time the ghost in the arcade was from, and also what kind of time the arcade itself was in. If the ghost disrupts the time and space of the arcade, then this is because the arcade now houses an undisclosed injustice – a death, a haunting: a tragedy for which there is not yet a history. Ghosts are witnesses to the losses and traumas that might disturb the free flow of time. Just like Klee's painting of the *Angelus Novus*, ghosts are spread-eagled by the catastrophe of progress. Of the painting, Benjamin famously observes:

> This is how one pictures the angel of history. His face is turned towards the past. Where we perceive a chain of events, he sees one single catastrophe which keeps piling wreckage upon wreckage and hurls it in front of his feet. The angel would like to stay, awaken the dead, and make whole what has been smashed. But a storm is blowing in Paradise; it has got caught in his wings with such violence that the angel can no longer close them. This storm irresistibly propels him into the future to which his back is turned, while the pile of debris before him grows skyward. This storm is what we call progress. (1940, page 249)

The ghost, like the *Angelus Novus*, is a proper witness to the catastrophe all about it. But the ghost is not trapped in the flow of time. Instead, the ghost exists in several times at once. It is of the past, wants something of the future, but is only connected to the present as an apparition, simultaneously visible and invisible. To be alert to the ghost – and the presence of ghosts – requires a particular kind of seeing. It is no accident that childhood provides the opportunity for uncanny effects in Benjamin's writings, for the paradox of innocence and cruelty crystallises in child-like ways of seeing. It is the contradiction between innocence and knowing, between belief and knowledge, and between the past and the future, embodied in the troubled child that underpins the film, *The Sixth Sense*. By discussing *The Sixth Sense*, it is possible to highlight the uncanny affect of living with ghosts in the modern city.

4.3 Seeing Dead People: childhood, the uncanny and living with loss

Starring Bruce Willis (as Dr Malcolm Crowe, a child psychologist) and Haley Joel Osment (as Cole Sear, a highly distressed young boy), *The Sixth Sense* (written and directed by M. Night Shyamalan) was one of the most successful films of 1999. This is one of those films that has the kind of plot twist that makes you want to replay the movie and rethink the entire story. Its popularity had much to do with the 'mysterious and unforeseen consequences' that the blurb on the back of the retail video-tape rightly talks about.

The film is a supernatural chiller, focused on Dr Crowe's attempts to learn what is disturbing the boy. To begin with, Dr Crowe has to win Cole's trust, and then figure out a way to exorcise his ghosts. In the final few scenes, the real ghost is revealed – both to the audience and also to the ghost itself. There are many creepy moments in the movie, usually playing on the uncanny affect readily associated with haunted children and haunted children's spaces and times.[19] This observation supports one of Freud's explanations for uncanny effects in literature.

Freud (1919) famously defines the uncanny as a feeling of horror and dread evoked when something familiar becomes disturbingly strange, creepy, fearful, scary. Freud explains:

> [...] the uncanny is something which is secretly familiar, which has undergone repression and then returned from it, and that everything that is uncanny fulfils this condition. (page 368)

This definition is not intended to cover all classes of horror, but it does describe many of those moments that make the hair on the back of the neck stand on end. Though Freud lists various kinds of experience that produce uncanny effects – such as doubling, dividing and interchanging people, repetitions, helplessness, the denial of death, the undermining of self-determination, evil and magic, and so on (pages 339–363) – he argues that many people experience uncanny feelings most intensely 'in relation to death and dead bodies, to the return of the dead, and to spirits and ghosts' (page 364). Thus, in *The Sixth Sense*, the familiar, sweet child becomes the unsettling medium through which puzzling and uncanny events occur – in the form of the ghostly manifestations.

For Freud, the essence of the uncanny is that it is caused by the return of the repressed. Repression occurs when traumatic experiences or painful thoughts are kept away from consciousness by various means. These repressed ideas and thoughts are then quarantined within the unconscious. If the opportunity occurs, however, these thoughts will erupt into consciousness – sometimes as an intense feeling, sometimes as a mood, sometimes in dreams, but only rarely as an explicit thought or memory. To return, the repressed takes advantage of a gap, however temporary, in the defence mechanisms that an individual uses to keep unconscious memories and feelings at bay. Significantly, according to Freud, this means that the uncanny experiences rely on fears and anxieties that are already known to the individual, but not necessarily consciously so:

> [...] this uncanny is in reality nothing new or alien, but something which is familiar and old-established in the mind and which has become alienated from it only through the process of repression. (1919, page 363–364)

In some ways, the uncanny has a dream-like, or nightmarish, quality. In it, feelings that have been displaced suddenly find a route – via some trigger in the waking world – to come to consciousness, not as an explicit thought, but through sensations that are displaced or condensed on to representative images. The main representative image in *The Sixth Sense* is the small boy, Cole Sear.

The figure of the small boy condenses many possible fears. Cole has knowledge beyond his years, he can see things adults cannot; yet he is unable to articulate his fears and lives in a terrifying world of his own, he is unable to secure aid from the adult world, he is literally helpless. Cole thereby embodies many of the fears and anxieties commonly experienced in childhood: loneliness, alienation from the adult world, an inability to communicate or be taken seriously, abandonment and separation. Cole himself becomes a creepy figure, embodying both intense childhood anxieties and also fearful contact with the dead.

From the beginning of the movie, the plot twist seems to involve the question of whether Cole (the child) can see ghosts or not and, therefore, whether he is suffering from a mixture of visual hallucinations, paranoia, schizophrenia and other kinds of emotional disorder. Though Cole seems to have every psychological disorder under the sun, the potential accuracy of the diagnosis is emphasised by Dr Crowe's status as an award-winning child psychologist. The question whether Cole Sear has the ability to see dead people[20] is, thus, dissipated. Even so, not only do we see ghosts as Cole sees them, but we also witness the physical impact of ghosts on Cole's body. So, if Cole can see ghosts, then the question becomes whether Dr Crowe is himself a ghost – and this is the central concern of the plot twist.

Amazingly, the film makes it very clear that the ghost is indeed Dr Crowe. The clue is transparent (perhaps too transparent, the film-makers worried). As a result of ending up in hospital, traumatised, Cole chooses to tell Dr Crowe his dark secret. Cole confesses that he sees dead people, and that these ghosts walk around like regular people, not realising they are dead because they see only what they want to see. Throughout Cole's startling revelation, the camera fixes unblinkingly on the face of Dr Crowe. The message is unambiguous: Dr Crowe is a ghost, but he does not (yet) know it because he sees only what he wants to see.

Let us see what *The Sixth Sense* shows us about the uncanny presences of the city. To begin with, the film is deliberately situated in a city – not just any old city, but a particularly old American city: Philadelphia. A caption informs the audience that, in a gesture that works as much to locate as to authenticate, the story will unfold in South Philadelphia. We see Dr Crowe going over some case notes about a patient with severe emotional problems. The street is quiet, and cold, there are very few people around. Exactly as we would expect, no-one is paying any attention to Dr Crowe. Then, as Cole leaves his house, running, frightened, to a church, Dr Crowe follows. We are not surprised when no-one takes any notice, either of the boy or the man following him. The streets are empty of both people and traffic. The church itself is also empty, even as Dr Crowe makes his first contact with the haunted child.

Throughout the film, the street scenes appear cold and empty. In this sense, there are no 'eyes on the streets', contrary to Jane Jacobs' wishes (1961): the contemporary American city is dead and deathly. There is no-one to, or who would, notice – let alone stop – a stranger following a frightened boy;[21] or, more in keeping with the film's reality, no-one to help a small boy apparently talking to himself, as if deep in a conversation with someone else. The streets, far from being places of mixing and meeting between people, are evacuated of human contact – evacuated, that is, except, ironically, for that between the living and the dead.

The choice of Philadelphia as the location for the story is no accident. One of the characters in the film, the schoolteacher 'Stuttering Stan' explains:

> Philadelphia is one of the oldest cities in the country. A lot of its generations have lived here and died here. Almost any place you go in this city has a history and a story behind it. Even this school and the grounds it sits on.

He goes on to ask whether anyone can guess what the school buildings were 100 years ago. Cole answers that they hanged people there. The teacher gets annoyed. He contradicts Cole: this was a place where laws were made, where law-makers gathered, where the very rule of law began in America. Cole is adamant: it was a courthouse, in which people were tried and hanged. The point is clear: there are many stories behind any place you go in the city – and some of those stories are about violence and death, even at the hands of the law-makers.[22] Cities, then, are places – and have places within them – that have a shadowy density about them. Visually, the film itself makes great use of shadows to convey this other-worldliness; this underlay of death, grief and trauma that pervades – yet is *actively forgotten* in – the course of city life. Indeed, some of the ways in which Dr Crowe is *misrecognised* as being alive have to do with the *amnesias* associated with the deathliness of city interactions.

It is not surprising that Dr Crowe can travel by bus, or attend a funeral, and yet not have to interact with anyone – because the city's unnoticed, unremembered, unrecognised strangers are everywhere. Similarly, Dr Crowe is placed alongside living characters, such as Cole's mother and Dr Crowe's grieving wife, to suggest that he is just as alive as they are. This is an effective demonstration of how the mere juxtaposition of people (and other things) is enough to convey interaction between them – as in a dream.[23] What this suggests is that city life is made and remade in the context of an everyday forgetting and the ordinary misrecognition of people by people. So familiar is this that even Dr Crowe himself does not realise that he is not in contact with the world. Now, it would be easy to think that this shallowness of interaction – the studied indifference and cynicism of the urban dweller – was a bad thing, yet what becomes apparent in this story is that these are also part of the ways in which people cope with the troubles and griefs in their lives and in the lives of those around them. Because, and this is the problem, in any city the reminders of death and strife are all around, ever present.

For Derrida, ghosts begin by returning (1993, pages 10–11). For him, this is what ghosts do: they return (from the past). His analysis is based on a reading of *Hamlet* – in which Hamlet's father has returned to finish the business of his murder. In the case of

The Sixth Sense, ghosts also return for help from the living; to help settle their unfinished business. However, other ghosts do not return, they haunt or possess (a location). What they want from the living may have more to do with their desire to haunt and possess a site, and with what it means to haunt and possess, than with returning to make demands of the living. In this light, Derrida's model of haunting and the ghost is based far too squarely on Hamlet's dad. That ghosts haunt and posses particular sites also raises the question 'what do ghosts want?', as we will see in the next section.

4.4 Building Sites: sighting ghosts in New Orleans, London, Singapore

In the previous two sections, there is a sense that ghosts are strangely familiar presences in the phantasmagorias of modern city life. What is distinctive about urban haunting – or haunting the city – is the sheer quantity, heterogeneity and density of ghosts. This implies that cities are places where innumerable pasts – tragic, traumatic or otherwise – co-exist. Each ghost, all ghosts, speak to the co-existence of these different events, each with its own temporalities running through it. Each site is possessed, haunted, by these times. Ghosts not only embody the heterogeneous temporalities of the city, they also disturb senses of place. Certain sites seem to have more ghosts than others: for example, cemeteries, but also theatres, derelict buildings, and sites associated with death, especially hospitals and locations of murder, suicide or execution.[24] Ghosts also possess places, often taking possession of domestic spaces, such as, famously, the haunted house. Such stories will expose both the heterogeneous temporalities that comprise places, but also illuminate what it means to possess, or haunt, a place.

In this section, I will look at three haunted domestic buildings – a house, a block of flats, and a modern apartment block – through three urban ghost stories: the first is Julie's, which is set in New Orleans; the next involves John Constantine, in London;[25] the last is Chang Kat's experience in Singapore. In each of these stories, the past holds on to the present. The ghosts demand something of the living and, in doing so, interrupt the living's relationship to both the past and the present. To begin with, let us hear Julie's story.

In Chapter 3, we heard about New Orleans' vampire tours. Alongside these are ghosts tours, and one favourite site is 732 Royal Street (see Figure 4.7). The story goes something like this. From the late eighteenth-century through to the American Civil War, it was customary for young white Creole (i.e. French-descended) men to take women of colour as their mistresses. The men often provided the women with their own homes and they would live relatively free lives.[26] The plaçage system meant that children born to these mistresses, though illegitimate and classified as black, would bear the name of the father and be supported by him. Sometimes the boys would be sent to Paris to be educated, but the girls would likely end up as mistresses of the next generation of Creole men. In the racial and racist accountancy of blood, these women were known as quadroons (one quarter African blood).[27] Indeed, between 1805 and the Civil War, the quadroon balls – where white men would openly select a quadroon mistress – became a significant feature of the New Orleans social calendar. The story of 732 Royal Street concerns one such quadroon, Julie. In the tales, this background – and its foundation in racial and sexual exploitation – is never mentioned. Instead, Julie's ghost is told as a tragic love story.[28]

On cold December nights on the uppermost balcony of 732 Royal, witnesses have seen the ghostly figure of a beautiful, naked, 'olive-skinned' (sic) woman, silently shivering in the cold, pacing up and down.[29] Sometimes other descriptions for her 'olive-skin' are used, sometimes she's 'nearly naked', but you get the picture. People on the tours, at this point, often scrutinise the building, hoping (fearing?) to see a naked beauty. Some shudder at the thought. One popular version of the tale continues like this.[30] Julie was the mistress of a wealthy young Frenchman. The unnamed Frenchman was keeping Julie, still the pair were deeply in love. But love across the racial divide was forbidden. One night, the lovers were surprised by a visit from his family. 'Don't worry', he said, 'I'll tell them about us and we will live happily ever after.' Or words to that effect. 'Look', he continued, 'I'll go down and explain everything to them, but you hide on the balcony until I say it is safe to come out.' Julie, though (nearly) naked and (always) beautiful, agreed. On to the balcony she went, on a freezing cold night in December in the year 17 or 18-something. She waited and waited. And waited. Never doubting that her nameless lover would eventually come to fetch her. Never doubting that what she wished for would come true. Desperate to prove her love, she quietly waited. Until she quietly died.

Figure 4.7
Julie is seen on the upper balcony at 732 Royal Street, New Orleans.

New Orleans' brutal racial history is presented in a sentimental light. Blame for the tragedy of Julie's death is placed on the failings of a wealthy young Creole. Social drama is rendered as a family drama: melodrama is rendered as a ghost story. In other New Orleans ghost stories, the slaves are not quite so doomed by their love for the white man. Ghosts of a plantation owner's family stalk one mansion, their fate was to be poisoned – allegedly – by one of their female slaves. Clearly, the barbarism of the slave system provided enough tragedy to leave a long trail of ghosts, all willing to display – if not exactly detail – the exact nature of the trauma.[31]

London is well known for its ghosts. Indeed, noisy matter-of-fact London can barely contain its many traumatic histories.[32] Most of these tales refer, as you might expect, to tragedies in the past, often involving suicidal aristocrats, horrific murders and deadly plagues. Taken together, London assumes a ghostly air, as if still cloaked in its infamous nineteenth

Figure 4.8
A watery grave? The Thames, London.

and early twentieth century smogs. For artist Roni Horn (2000), even the surface of the River Thames has a haunting quality (Figure 4.8), as its unfathomable depths recall the dead bodies that have been found there.[33] Also haunted are London's tower blocks, including its high-rise social housing. In Neil Gaiman's tale, 'Hold Me' (1995), John Constantine finds out just how.

In the tale, John Constantine is making the journey to the wake of his friend, Ray. On the way, he ends up walking through one of London's poorer neighbourhoods. The scene is set: it's a place of drunks, beggars, dirt, poverty, and squalid social housing. At the wake, John meets Anthea. Anthea tells John she's feeling unwell, so he escorts her back to her flat, in a nearby tower block. The tower block is stereotypical: the lifts smell of urine, there's graffiti on the walls, it's desolate, and it's cheap. But tonight the smell is unusually bad, as if a cat had died in the next-door flat, number 512. Not long after getting to Anthea's flat, there's a knock at the door. A distressed child tells Constantine that her mommie is dead. Ruefully, Constantine promises to investigate. He breaks into flat 512. There he is confronted by the sight of a dirty, scruffy, old – and dead – man. The dead man grunts 'hold me' at Constantine (page 20). Constantine recoils in disgust. He thinks to himself: 'Oh great. Dead of the living night. Night of the living dead … whatever. What am I *doing*? Think.' (page 21). The dead man mutters incomprehensibly, something about holding him perhaps?

Initially, John cannot tell what the dead man wants, nor can he decide what to do. Then, Constantine opts to hold the dead man, as this is what the dead man seems to want. 'You're freezing cold and you smell like a rotten abattoir, mate', Constantine observes, 'Must be hell being dead' (page 22). And then the ghost disappears. The ghost, in this story (and not as Derrida would have it), begins by waiting (for something). However, what is important is that the tower block is a site for haunting because poverty itself is a tragedy, and it kills – that is, because poverty haunts London, one of the richest cities in the world. New buildings in old cities can, under frightful circumstances, become sites of haunting.

What is true in London is also true in Singapore, but differently so (Figure 4.9). There are a number of ghost stories associated with tower

Figure 4.9
A block of flats, Singapore.

blocks: for example, Lee recounts a story reported in the reputable *The Straits Times* on 10 March 1985, concerning the sighting of a ghost in a block of flats in Bukit Batok (1989, pages 66–68).[34] Another tale concerns a young urban professional, Chang Kat (Anon, 2000b). The story is told by a friend and it begins like this (page 119):

> 'Hey, Chang Kat! I heard that you are buying an apartment in Johor Bahru right?'
> 'Yes, it's really a good bargain. Where else can you buy an apartment at such a price? It is a steal.'
> 'True, but I heard that the area is "dirty". People always tell me that they had seen pontianak and ghosts around there. Are you sure you want to live there?'
> 'You read too many Nightmares books. Pontianak and demons normally hang around old places, buildings and graveyards. New apartments are not for them.'

Chang Kat is not to be put off a bargain apartment by superstition. In this, Chang Kat is unlike many Singaporeans. For Chang Kat's friend, the living world is not the real world. Instead, what westerners – the archetypal modern city folk – call the other-worldly *is* the real world. The real city, for him, is other-worldly. So, any decisions made in the unreal living world would do well to pay attention to the real other-world. This is, it is widely believed in Singapore, reflected in house prices. Thus, house prices are supposed to fall in the seventh lunar month, when the hungry ghosts are about, as this is an ill-fated time to sell/buy houses. Anyway, Chang Kat is unconcerned about the low price of the flat. He should have been.

One day, Chang Kat arrives home from work to find a beautiful woman, with luscious red lips and wearing a short white translucent dress, standing in the garden next to his apartment block. She says 'hello', and he responds in a friendly way, thinking his luck is in. As he gets to the building's lift, he finds her blocking his way. Amazed at how fast she has got there, he barely notices that he has just walked straight through her into the lift. For a second, Chang Kat freezes. Then he runs terrified to his flat on the ninth floor. There she is again, but now she's *in* his flat. 'Am I beautiful?', she purrs. Chang Kat is too terrified to answer: realising the woman is a ghost, he is dumb-struck. He closes his eyes and prays for deliverance. She laughs. Then, Chang feels a sharp pain in his back, as she digs in her claws. Fortunately, Chang's screams wake his father, who bravely chases the ghostly apparition away (with a broom).

What are these ghosts doing in these buildings in New Orleans, London and Singapore? One answer to this question can be found in Sharon Marcus's (1999) explanation for 'haunted house' stories in Victorian Britain, many of which were located in London.[35] She explains the prevalence of haunted house stories primarily through changing property relations. At the time, many houses were overcrowded, houses were being subdivided into smaller units, and the inhabitants of such dwellings moved on after short periods. Fears surrounding not knowing who lives next door, not knowing who lived in the house before, and not knowing what might have happened in the house are expressed – and, where exorcisms are successful, allayed – in haunted house stories.

For Marcus, the physical characteristics of the Victorian cities also lend themselves to ghost stories. Cities such as Paris and London had narrow, ill-lit, labyrinthine streets, secret spaces, ruins, catacombs and tombs.[36] Victorian houses, meanwhile, had marginal, secret spaces, such as basements and attics. In these, who knows what might have happened? During restoration work on Benjamin Franklin's house (at 36 Craven Street in London), for example, more than 1,200 pieces of human bone were discovered in the basement.[37] Dark deeds in Franklin's house? Or simply innocent medical practices? Not just closets have skeletons.

Ghosts, in Marcus's account, do 'double duty'. They express anxieties about property relations in the city, while at the same time allaying these anxieties through exorcisms of various kinds. This is, in effect, what happens in the John Constantine story. Anxieties associated with the possibility of living in a place where an old alcoholic has died, alone and unnoticed, are allayed by Constantine's sympathetic and understanding exorcism. The dead man was only seeking comfort and all that was required was an appreciation of what the ghost wanted.

Through ghosts, the past comes back to unsettle the living, who often arrogantly assume that they have sole possession of a place. Julie is a silent reminder of the colonial and slave pasts that still live on in New Orleans. People are most caught out when they either do not know the history of a place or when they decide to ignore the warning signs. The consequences of the possession of the present by the past can be grim. Chang Kat narrowly escaped with his life. John Constantine, meanwhile, discovered that mommie had indeed been killed by the old man's ghost.

The past is not always easily exorcised. Julie still appears on the balcony, forever waiting to be let in – an involuntary memory of the callous cruelty of slavery. What Chang Kat did not appreciate is that many of Singapore's Chinese cemeteries have been sold off for housing developments. This means that, not only are there worries about who might have lived in a flat before or what might have happened there, there are also anxieties about whether the foundations of buildings might be directly connected to the dead (see Figure 4.10). This is why there might well be a ghost waiting in the dead space outside a block of flats. Marcus suggests that ghosts provide an accurate picture of bourgeois life in Victorian London and Paris. Similarly, ghosts speak to the haunting quality of contemporary life in historic New Orleans, in squalid London and in modern Singapore.

Figure 4.10
Cities of the Dead, too: a Chinese cemetery in Singapore.

Ghosts are disruptive figures in the phantasmagorias of time and space in New Orleans, Singapore and London. In another *Hellblazer* tale, this time written by Warren Ellis (1999), John Constantine reflects on what it means to live in a haunted city like London:

> London's built on ghosts. You can't walk London without walking on ghosts. Downriver, there's the row of council houses where razor blades fly through the toilets and living rooms [...] The Smithfield Meat Market, where they executed William Wallace for being an awkward Scotch twat, stinks of it. Literally. They boiled a cook alive there in the 1600s for poisoning some Bishop's mates. On a good night, you can smell him cooking, him and other heretics and turbulent priests they roasted here. (pages 98–99, 100)

The meat market was consolidated on the Smithfield site, one of the bloodiest places in London, as the result of nineteenth-century reforms.[38] It is certainly a place of blood and bones. And also, as tour guides will tell you, of ghosts. An alternative map of London would show its histories of blood and bones and ghosts. London's present treads upon these pasts, these heterogenous histories. Each site, each haunted house, is associated with heterogeneous temporalities. Not the smooth flow of a singular history, but the fractured and fragmented times of traumatic events as they seek to possess and haunt the present.

London is haunted both by ghosts (the dead) and by John Constantine (the living). Haunting has a double meaning: it refers to the occupation of a place both by the dead and by the living. The ghost-like figures that appear in the phantasmagorias of city life, therefore, can be either dead or alive. But Constantine makes one further observation, with a bitter smile on his face: 'My name's John Constantine, and here I stay: haunted by London. And London, haunted by me' (page 142). Thanks to John Constantine's cynical and bitter observation, we can now see that he himself is haunted by London. The city is also haunting. The city itself is spectral presence, ghostly – a trauma unto itself. I will develop this aspect of cities in the next section.

4.5 New York Dead and Buried: the city and grief-work

In most ghost stories, how the dead got to be dead is crucial. Sometimes, the living are haunted by the guilt associated with the death: ghosts, the spectral embodiment of the feelings that the living have towards the dead (see above). For Freud, death presents the living with the difficulty of how to grieve appropriately. Freud suggests that there are two alternative forms of **grief-work**: mourning or melancholia. In mourning, the living recognise that the dead have departed. Slowly, the living learn to live with their loss and their broken emotional attachments to the dead. In melancholia, however, it is as if a part of the self had died. According to Freud, the melancholic experiences

> a profoundly painful dejection, cessation of interest in the outside world, loss of capacity to love, inhibition of all activity, and a lowering of the self-regarding feelings to a degree that finds utterance in self-reproaches and self-revilings, and culminates in a delusional expectation of punishment. (1917b, page 252)

To exemplify this form of melancholia, and also to locate it firmly in the city, I will examine Joe Connelly's novel *Bringing out the Dead* (1998).[39] In the book, Connelly vividly describes two moments when the dead are present in the city. These experiences are based on his own, when Connelly spent nine months working at St Clare's Hospital in New York City. The story he tells in the book concerns Frank Pierce, who is working as a paramedic at *Our Lady of Mercy* Emergency Room in New York's Hell's Kitchen. Frank is on the edge of a breakdown, running on alcohol and pills to keep at bay his feelings of grief, dread, fear, trauma and guilt; he is also seeing ghosts. Frank, moreover, is ready to give up the ghost precisely because he is so haunted; haunted in particular by Rose, the ghost of someone he believes that he has killed. Early on in the book, Frank's relationship to the dead is graphically presented. He and his partner, Larry, attend an emergency to find an old man, apparently dead, surrounded by his distressed family. The two paramedics immediately prepare to revive Mr Burke. As Frank opens Mr Burke's mouth, he feels a cool breath of air on his fingers:

> I'm sure it was just gas built up in the stomach from CPR, but in the last year I had come to believe in such things as spirits leaving the body and not wanting to be put back, spirits angry at the awkward places death had left them, and though I understood how crazy I was to think this way, I was also convinced that if I looked up at that moment I would see the dead man standing at the window, staring out over the tar-paper plots and gray ditches of his birthplace. (page 3)

The sense that the dead haunt the living pervades the story. In the end, Frank will ease the departure of Mr Burke's spirit from his body, so that his spirit will not be angry and resentful at being kept in the intense pain of intensive care. However, Frank's belief in spirits begins with his failure to save Rose. Her death haunts every other death: it haunts every emergency Frank attends; Rose haunts Frank, and she haunts the city; the city also haunts Frank. As he says:

> These spirits were part of the job. I always knew that. I had worked Hell's Kitchen for years, and it was impossible to pass a building that didn't hold the spirit of something: the eyes of an unloved corpse, the screams for some loved one. In the violent eruption of life that is death, all bodies leave their mark. You cannot be near the new dead without feeling it. I have seen even the quietest night passing fill a room the next morning with a dense fog of life's remains, and after the air has

cleared, the mark lingers, in a pillow that never regains its shape or a stain on the wall that never comes clean. (page 54)

The spirits of the dead become a part of the very fabric of the city, woven into its physicality, such that the sidewalks and stairwells become spectral, attached, in some way, to the spirits of the spirits – like a shadow cast by the body of reality. More than this, the living themselves carry around with them a double hauntedness: for they are also the not-yet-dead, yet also always in contact with death. The real horror is that, in the ghostly phantasmagorias of city life, the living become ultimately indistinguishable from the dead. As the story unfolds, we witness the torment that results from this (never less than) double aspect of the city – material and spectral; living and dead; present and absent; traumatised and numb; fearful and indifferent – of life in Frank's torn world. While driving around looking for a patient, Frank glimpses a ghost:

> I couldn't see her face, but I knew the yellow raincoat she wore, the black stockings with a hole in the knee where she fell down once because she couldn't breathe. It was Rose, the girl I had helped to kill. We all make mistakes, but some things cannot be forgiven [...] Rose was getting closer. Maybe it wasn't Rose [...] I crossed my arms on the steering wheel, and holding my breath I pressed my forehead on my hands until they were numb. One of the first things you learn on the job is how to block out the bad calls. In the same way that cops fence off a murder scene, you wall these memories up in the deepest cave in your brain. I used to be an expert, but lately I'd found some big holes. (page 53)

Far from being safely secured in the caves of the mind, memories – bad memories – are perfectly capable of finding their way into daylight, past the police lines. There have to be escape routes for them, however. The sight of a yellow raincoat; a faceless girl; a street scene; a sound – almost anything could allow the ghost's safe passage out of the deep, walled-up cave. The spirits are angry enough to keep returning, for returning is what ghosts most certainly do. Apologies don't appease these ghosts: they won't stop coming back.

Indelibly, the city bears the traces of its traumas. For Freud, memory-traces are indestructible, sunk into the deepest parts of the mind.[40] In this tale, the city is like the mind: forever bearing the memory-traces of the dead.[41] The city, moreover, cannot be cleansed of the remains of the dead. If it isn't bodies that leave their mark, then it is the escaping spirit that occupies a room, looking back on a life departed. And more:

> What haunted me now was more savage: spirits born half finished, homicides, suicides, over-doses, and all the other victims, innocent or not, still grasping at lives so abruptly taken away. Rose's ghost was only the last and most visible of many who seemed to have come back solely to accuse me – of living and knowing, of being present at their deaths, as if I had witnessed an obscene humiliation for which they could never forgive me. (page 55)

Frank is haunted, not just by Rose, but by the angry accusations that ghosts make against the living. In Freud's discussion of ghosts, he describes how a belief in ghosts is associated with the idea that the dead are angry because they are no longer living. For Freud, ghosts represent the living's ambivalent relationship to the dead. He describes the kinds of ritual practice (associated mainly with disposal of the body and mourning) that are involved in keeping the dead at a distance and in appeasing the dead precisely so that they do not return to haunt the living.[42] For Frank, in his melancholic state, there is no reconciliation with the dead. In New York City, the infrastructures of death are at breaking point: everywhere the ghosts of the dead accuse the living, everywhere death is an obscene humiliation, an injustice. Yet Frank *knows* he is not well, but the certain modern knowledge that ghosts are just tricks of the mind is not enough to exorcise the trauma and pain:

> It was a sickness, of course, the mind playing tricks. That's what I told myself, for that was the only way out. I'd been down before and always come back, but never before had I been so closely tied to someone's death. There were times when Rose could look as real to me as Larry. (page 55)

The Sixth Sense's Cole, then, is not the only person who sees dead people wandering around the city like regular people. Like Cole, Frank sees the dead standing at the window – looking in, accusingly, on life. But Frank also is haunted by the city itself: by the souls that fill the very foundations of the city; by its traumas; by the remembrance of times past. For both Cole and Frank, though, the city is haunting because it gathers together so many ghosts: there are so many reasons to be haunted, so many accusations against the living. In these tales of ghosts and the city, grief explodes like a bomb that shatters the immediate physicality of the city, that sends shock waves through the living, that leaves the traces of bodies in the softest and hardest parts of the city: in its pillows and in its buildings. The problem of ghosts, then, is less about the spectral return of the dead and more about redressing the traumas and injustices of the past. But, in the face of such a crowd of ghosts, care also must be taken of the living – and the traumas and injustices that they face.

So, Avery Gordon (quoted at the beginning of this chapter) is right to suggest that the ghost is a social figure through which something lost or barely visible can be made to appear in front of our eyes. And it is possible to read *Bringing out the Dead* in this light: through a sense of the loss of love or caring or understanding or recognition or reparation or forgiveness in the city; through the deathliness of life on the streets; through the ordinary indifference of city life (as Simmel would have it). Gordon is right, also, to point out that ghostliness is an epistemology: a way of coming to know the traumas and violence that accompany life, even while they are not acknowledged (repressed or oppressed). The ghost, finally, is certainly a way of bringing back these stories, of making the injustices of life walk among the living – a way of calling for justice. However, the ghost also presents a problem for the living. For, if the dead overshadow the lives of the living, if the living are always only ever in debt to the dead, then there is a danger in living only according to the dead projects of ghosts – in a perpetual melancholia where the only proper concerns of the living circulate around death.

4.6 Spooks in London: melancholia and anti-colonial struggle

According to anti-terrorist officers, at 10.46pm on Wednesday 20 September 2000, there were reports of a blast at Vauxhall Cross, on the south bank of the Thames, in the

Figure 4.11 On 20 September 2000, rocket-propelled grenades struck this building around about here.

vicinity of a large riverside office-block.[43] Other official police sources indicated that the blast (only one) could have been a bomb. Informally, ordinary police officers were pointing to damage high up on a prominent building (Figure 4.11) and, not unreasonably, speculating that something might have been fired at, and hit, it.

The news media (press and television) were less circumspect about what had happened and to whom. After the official statements, they reported eye-witness statements. While there was not agreement about the time – some putting it as early as 9.45pm, others certain it was more like 9.55pm – they all agreed that there were two explosions: the sound of one running into the other. Photographs of the damage were included: a small hole could clearly be seen below an office window on the eighth floor of the building. The building is now clearly recognisable as 85 Albert Embankment, Vauxhall Cross. The conclusion was

Figure 4.12 MI6 'Secret Intelligence Service' Headquarters at 85 Albert Embankment, Vauxhall Cross.

clear: there had been a terrorist attack on the building. It was speculated that terrorists had used a 'barrack buster' mortar to fire two missiles. It was added, as a matter of fact, that the target was the headquarters of the British Secret Intelligence Service, MI6.[44] It was popularly concluded that the 'Real IRA' (Irish Republican Army) had attacked 'MI6 headquarters': an obvious target (see Figure 4.12).

It matters, of course, which secret intelligence service was attacked. It could have been MI5. MI5 is tasked with countering threats to national security: their remit has been extended since the end of the Cold War, beyond counter-espionage, to include combating 'terrorism' (mainly the IRA), the drug trade and other domestic threats. However, this was not their headquarters. The building, matter of fact, is home to MI6. Now, if MI5 is about spy-catching and counter-terrorism, then MI6 is about spying – not something governments generally own up to, and the British Government is no exception. MI6 agents are recruited to infiltrate foreign governments and groups (including terrorist organisations, such as Al-Qaeda) that are considered a threat to British national security. Press reports, significantly, only mentioned MI6's role, alongside MI5, in countering Irish terrorism – thereby casting a shadow over MI6's spying activities *outside* the United Kingdom.

The shock attack temporarily threw a bright light on two secret organisations (the IRA and MI6), both involved in shadowy violent conflict. Out of the shadows, what could now be seen was the life-after-death of two ghosts, still slugging it out. A dead and deathly colonial war was being fought – in the heart of London, within sight of the Houses of Parliament. Ghosts? Fighting a dead war? After the Cold War, after the Northern Ireland peace process, we might imagine this particular anti-colonial struggle would be an anachronism. The attack, in this sense, was a real blast from the past. So, here we are – at the intersection of past, present and future – in the presence of two ghostly figures: the first embodied in the shadowy Irish Republican Army (whether provisional or real); the second comes under the popular name of 'spooks'.[45] First, let's look at the appearance (and disappearance) of the IRA.

Though the news media could not be sure whether it was the IRA, provisional or real, that had done it – since there had been no confirmation from any official source, IRA or not – the ghosts of IRA attacks past, present and future readily gathered at Vauxhall Cross. Recent attacks within London were cited as 'proof' that this was the work of Irish dissidents. There was, for example, the (successfully detonated) bomb at Hammersmith Bridge in June 2000 and the (unexploded) bomb at Ealing Broadway tube station on 19 July 2000 (which disrupted celebrations of the Queen Mother's centenary). Unsurprisingly, fears were expressed that this marked the start of a fresh mainland bombing campaign, especially in the lead-up to Christmas. So, not only IRA ghosts became visible at the bomb site – the kinds of ghosts who can lay explosives, but are never seen, and are certainly never captured – but also the ghosts of IRA victims began to gather. For London has often been witness to the pain and grief of acts of terror. And London is haunted by those who have died, or been wounded, as in the most dreadful of recent attacks:

- the Hyde Park bombing of 20 July 1982 killed 11 soldiers and injured 50 others;
- the Harrods bomb killed six and wounded 90;
- a huge explosion rocked the City of London on 10 April 1992, killing three;
- the Bishopsgate attack of 24 April 1993, where a truck was used, killed one and injured 44;
- a bomb on Canary Wharf in London's prestigious Docklands on 6 February 1996 ended the provisional IRA's cease-fire by killing two people.

Twenty-three dead to add to the living ghosts of the (Real) IRA: 'ghosts', also because the peace process was supposed to kill off the armed resistance to British colonialism in Northern Ireland. Still these ghosts are real, the past will not be buried (even by the tortuous process of peace). Ghosts living and dead told of the pain and grief that cities – cities like London – hold close to their hearts. But this event also reveals much about city ways of life. For, alongside the terror and grief, there is also a story of ordinary indifference, of calm curiosity and of people choosing just to get on with their lives. Thus, the eye-witnesses seemed relatively untroubled: their reports were calm, many simply keen to see what had happened and to learn what had caused the two loud bangs that they had heard.[46]

By mid-November that year, stronger fears of yet another IRA Christmas bombing campaign were being expressed. This time, ironically, in the wake of better news from the security services. On Sunday 12 November, according to police sources, a plot to detonate a 500-pound bomb (somewhere) had been foiled. The huge bomb was made of nuts and bolts wrapped around an explosive core comprised of fertiliser. Once again, the spectre of the IRA had returned to London. The bomb was similar in size to the one that had rocked Canary Wharf and – despite the fact that it was also of similar size to the bomb used in Omagh, Ireland, that killed 29 people in 1998 – it was immediately assumed that the bomb was destined for a prestige target in London and, moreover, because of its size, that the IRA's intention was to kill as many people as possible. There was an agitation to London's ghosts now: explorations of events in the present were being ordered around the presumed motivations, methods and outcomes of past deeds – and these were, also, extrapolated into feared futures.

On Saturday 25 November, in news that both confirmed and allayed fears, the discovery of an arms cache in Londonderry was announced. The Sunday papers carried warnings of a terrorist offensive. By Monday, television news programmes were carrying the official message – be afraid. Londoners, meanwhile, were far more concerned about train derailments, delays on the tube and cruelty in its poorest housing estates (especially in the wake of a series of child murders). London has become accustomed to its ghosts, for it is home to more ghosts than those of its Irish colonial past. For, London has been witness to killings that have raised many other spectres from its imperial legacy. In recent years, the conflicts between Pakistan and India, between Kurds and Turks, between Palestine and Israel, between the Muslim world and the western world have led to violence on the streets of London – especially in the wake of the attacks on New York's World Trade Center and the wars in Afghanistan and Iraq.

Far from it being obvious that it was the Real IRA that had attacked MI6, London can gather in many other groups wishing to use it as a place to find justice for their dead. Perhaps others had spooked the spooks? After all, it is one of the worst kept secrets that MI6 spooks haunt the spectacular Thames-side office block. Thus, for example, to overcome a British Government ban on filming the exterior of Vauxhall Cross, James Bond is seen launching a power boat from an exact replica of MI6's HQ in *The World is Not Enough* (1999). The real Vauxhall Cross, meanwhile, appears in Patrick Keiller's film, *London* (1994),[47] and in an earlier Bond movie, *Goldeneye* (1995). Indeed, while it was being built, Vauxhall Cross attracted controversy: not simply because it was such an *obvious* place for the British *secret* intelligence service, but also because of its high cost. Much of this cost, it was speculated, paid for MI6's high security requirements, including anti-surveillance and bomb-proofing. These secret security added-extras neither deterred the Real IRA, nor did they help catch the attackers. So much for Mr Bond.

Indeed, as far as Bond stories go, this wouldn't really make the grade: there were no heroes and no world-class bad guys got their come-uppance. Instead, the two sets of ghosts seem to have slid past each other, each with barely a scratch. Indeed, these two ghosts seem more concerned with appearances in public than with actually defeating one another. For the IRA, MI5 would logically be the better target militarily. After all, it is MI5 that is most directly involved in anti-terrorist operations against the IRA. However, their offices are more scattered and less visible and so more secret and, therefore, less likely to gain much publicity. The MI6 building stands out like a sore thumb. It was an obvious propaganda target – an unmissable office-block, at night lit up like an elaborate Christmas cake. How could they miss? Ghosts are, as we can see, all about show. Ghosts are also about disappearance: the spooks, in this story, were never really there. MI6 are not visibly there, so the attack was launched at no-one – by no-one.[48] Suddenly, the real (whether official or IRA or physical) shivered as if a ghost had drifted across the city, before our very eyes.

A procession of ghost-like figures – figures who were neither fully present, nor fully absent – walked through the aftermath of the rocket attack. Among them were terrorists, freedom-fighters, victims and bystanders; 'foreigners', locals and Londoners; spies, the police and 'official sources'; commuters, eye-witnesses and residents; news reporters, TV cameras and photographers – and, through them, yet more observers. Noisily, and matter-of-factually, many of these ghost-like figures speculated, a few reasoned, some clarified, others obfuscated. The terrorists remained silent, floating ominously through the walls of the building and the event itself. As these ghosts passed back and forth, Vauxhall Cross was becoming overdetermined – both with meaning and with emotions: alarm and surprise, confusion and conjecture, laughter and indifference;[49] 85 Albert Embankment shook with meaning and emotional intensity, as if someone had stepped on its grave.

The assault on 85 Albert Embankment betrays how cities become dense sites of history and subjectivity, where people make public their political demands, symbolically and sometimes violently. It also tells us something of the ghosts that violent events set loose on the world, and the ways these ghosts intensify the feelings and emotions which underpin understandings of those events. These ghosts, moreover, do not simply go away once the catastrophe is over: they hang around, haunting the places where they made their appearance. Ghosts, then, live-after, among us, all the time. Cities, in this light, are places where diverse ghosts, given the opportunity, can gather in large numbers.

In cities, grief-work – whether mourning or melancholia – is continually being conducted. London does not just house the dead, its dead crowd its streets. Cities are places where ghosts can gather, uncannily, spookily. They are also places of mourning and forgetting, where the ordinary violences of everyday life are simply lived through; where few give ghosts much attention, and not for long when they do. Simmel caught something of this in his study of the city and its mentality of indifference (1903).[50] However, we can also perceive aspects of this in the accounts of the attack on MI6. The newspapers were full of stories from bystanders, revealing people's indifference to trauma and their feelings of safety; something also of people's curiosity and their willingness to *see* something that could not be *seen*; something too of *schadenfreude*, of the desire to witness the trauma, without experiencing its consequences. Disturbing events are witnessed, rationalised emotionally, then forgotten.

Even so, the ghosts – or clash of temporalities, if you prefer – embodied in the attack on the MI6 building suggest that melancholia is part of the phantasmagorias, and the emotional life, of city life. The dead – the ghostly markers of loss, trauma and injustice – are sometimes almost impossible to give up. The lesson of the attack, alternatively, is that the dead must be laid to rest. The living must get on with living, with finding new ways to live with one another, instead of trying to blast holes in one another. It seems as if getting rid of ghosts is a good thing. Exorcise them! Let the dead bury the dead, as someone once said. But is this the best way to deal with ghosts?

4.7 Nightmares of the Living: ghosts versus anti-capitalist protesters in London

For Marx,

> The social revolution of the nineteenth century can only create its poetry from the future, not from the past. It cannot begin its own work until it has sloughed off all its superstitious regard for the past. Earlier revolutions have needed world-historical reminiscences to deaden their awareness of their own content. In order to arrive at its own content the revolution of the nineteenth century must let the dead bury their dead. (1852, page 149)

For Derrida,

> It is necessary to speak *of the* ghost, indeed *to the* ghost and *with* it, from the moment that no ethics, no politics, whether revolutionary or not, seems possible and thinkable and *just* that does not recognize in its principle the respect for those others who are no longer or for those others who are not yet *there*, presently living, whether they are already dead or not yet born. (1993, page xix)

Politics, whether revolutionary or not, whether nineteenth-century or not, would seem to have to contend with the dead, with ghosts – well, according to Marx and Derrida, that is. If we're to believe them, then any story we might wish to tell about the free and fair city would have to take into account its ghosts. On one side, Marx would be there telling us that we have to be free of the spirits of the past; we revolutionaries must let the dead bury the dead. On the other side, Derrida counters that it is not so easy to exorcise our ghosts; instead, in the name of justice, for those who have died, who have not yet died, and who have not yet been born, we must speak to and with the ghost. If the city is to be a site of freedom, justice and democracy, then, there is a troubling problem: what is to be done with the tradition of dead generations that weighs so heavily on the lives of city dwellers? I will explore this question through the events surrounding the anti-capitalism protests around Parliament Square and Trafalgar Square on 1 May 2000 (unofficially Labour Day in the UK).[51] I draw out the emotional intensity surrounding these protests by showing how ghosts came to haunt political actions and reactions.

On 1 May 2000, 4,000[52] people gathered in London's Parliament Square (Figure 4.13) to protest against the injustices of contemporary capitalism. For many observers, this was a further example of an increasingly radical (and violent) turn in popular protest. It bore echoes of rioting that had occurred the previous year during protests in London's financial district on 18 June (known as J18), as well as events in other cities around the world,

Figure 4.13
Parliament Square, London, where protesters massed on 1 May 2000.

such as (notably) Seattle.[53] The police, for one, not wishing to be caught unprepared (as they had been at the J18 demonstration) had mobilised military fashion, with the biggest operation in the capital for 30 years. There were 5,500 officers at the event, with a further 9,000 officers in reserve: a total, then, of about 14,500 officers.[54] Even acknowledging the likely undercounting of the number of protesters (one would normally double official estimates), this would make as many police as protesters. Let us think about this: so, a bigger police operation than during the anti-nuclear, anti-war, trades union, feminist, anti-racism and anti-apartheid demonstrations of the 1970s and 1980s (and 1990s). Somehow, the brief, but explosive, violence of the J18 protests had put the police – and the authorities – on their guard. What kind of revolutionaries were these?

Protesters had been gathering around Parliament Square from about 10.00am onwards. Perhaps unsurprisingly, several hundred cyclists arrived first (London's public transport being what it is). Others followed. Many (political) groups began to gather: from anarchists and communists, to situationists, to ecologists and gardeners, and even a few tourists. Though the main 'disorganisation' at the heart of the demonstration was *Reclaim the Streets*, the diversity of groups was visible in the many flags and banners that festooned the streets. But particular prominence was paid, in the press, to the direct action known as 'guerrilla gardening': some described it as a typical English garden party, but without the tea!

By 11.00am, green-fingered activists were busy digging up the grassy area outside Parliament and turning it, rapidly, into a garden – planting seeds, vegetables and flowers – complete with ponds and the odd gnome. The turf from the diggings was taken and laid on the roads. Squares, roads and streets were being reclaimed. Even the smell of the city had changed: from that of carbon monoxide and tarmac to that of earth, turf and manure. There was music too: a samba band played, drummers went a-drumming, whistles were blown as at the Notting Hill carnival, and there was dancing in the streets. All a bit untidy, a bit chaotic, and perhaps a bit extreme or nonsensical, but somehow not that bad.

'Not that bad' was about to turn bad. Sometime around 1.40pm, some of the protesters began to attack various businesses in Whitehall, on the fringe of Trafalgar Square. A couple of young men were caught on

Figure 4.14
The McDonalds in Whitehall that was attacked by protesters on 1 May 2000.

CCTV breaking the front window of McDonalds (Figure 4.14); one wearing a black mask, gloves, and wielding a chair. Soon, the signature M sign was pulled down and destroyed; 'McExploit' had been sprayed in red on to the wall; 'McShit' in black. Next door, a *bureau de change* was broken into, as was a nearby souvenir store. Guerrilla gardening was turning by degrees into a set of attacks on properties that signified, in some way, all the evils of capitalism (in its global guises).

Now property was under threat, the police intervened. At around 2.30pm, the police acted in defence of law, order and, it must be added, capitalism in all its guises. In full paramilitary paraphernalia, baton-wielding, body-armoured ranks of police began smashing protesters' unarmed, unprotected bodies. Whether window-smashing anarchists or plant-wielding gardeners hardly mattered. It is not as if the police could distinguish between them anyway. Guerrilla gardening quickly became guerrilla warfare, with short battles being fought around Trafalgar Square for almost four hours. From the beginning, the police had blocked all possible escape routes and begun to squash the protesters into Trafalgar Square. By 6pm, the police (in their words) retook Trafalgar Square. Though sporadic street fighting continued into the evening (as the police pushed small groups of demonstrators up the Strand), civilisation and the city at last were saved and safe. Now was time for political comment.

Prime Minister Tony Blair weighed in with this:

The people responsible for the damage in London today are an absolute disgrace. Their actions have got nothing to do with convictions or beliefs and everything to do with mindless thuggery.[55]

Predictably, there was a failure on the part of parliamentarians to see the grace and mindfulness of the protests. And not just politicians, liberal political commentators bemoaned the lack of a proper political agenda in the demonstrations: heavy-weight journalist Hugo Young interpreted the actions as stemming from 'a bogus romance with anti-politics' (the *Guardian*, 2 May 2000, page 18). Hugo Young even underscored the futility of the 1 May protests by making unfavourable comparisons with the demonstrations in Seattle and Washington, which, he argued, had intensified political debate on the consequences of global capitalism – he concluded, to the benefit of consumers!

Now, such controversies would probably have arisen over any such event in London. Few had said good things about the Poll Tax riots, some ten years earlier. Nevertheless, in public debate over the day's events, the red and green revolutionaries were increasingly hidden by legions of ghosts, as they marched right through the May Day protests. Out of the wreckage of these few London streets, hundreds of thousands of the dead came to haunt the minds of the living. Let us watch the ghosts drag themselves up from their unmarked graves and join the phantasmagoria of politics (even if we cannot easily talk *with* them, as Derrida would wish).

Sometime between 2pm and 4pm, the Cenotaph in Whitehall had been defaced. The Cenotaph in Whitehall is Britain's monument to its war dead. Primarily, it is associated with the dead of the two world wars – 1914–18 and 1939–45 – although it is actually intended to commemorate all of 'the fallen', including those up to the present day. Every year, on a Sunday in early November, Remembrance Day is observed by royalty, political leaders, the armed services, veterans and relatives, when they gather at the Cenotaph to lay wreaths and pay their respects to the dead. A minute's silence is observed and then broken by buglers sounding the Last Post. Instead of the drums, whistles and dancing of J18 and 1 May, a silent national anthem plays over London's deadened streets. The monument itself is made simply of white stone, with a stark presence that itself calls to mind the grim finality of death (Figure 4.15). Perhaps this is how monuments to the dead haunt the streets of cities. They enable the dead to hold fast to the city, making sure that their time is not forgotten, not passed.[56] Even so, exactly how monuments break into the memories of the living can be unpredictable, contested and even shocking.[57]

Figure 4.15 'Our Glorious Dead', The Cenotaph, London.

For many, it was the desecration of this national symbol of sacrifice and grief that most starkly told of the 'mindlessness' of the demonstrators. Most offensive of the graffiti was this: on one side of

the ghostly city

the war memorial two green arrows, one marked 'MENS' and the other 'WOMENS', indicated where the 'TOILETS' were. The ultimate sacrifice of hundreds of thousands of (mainly) young men and (many) women had been profaned. On the evening's television news programmes, survivors of both world wars and relatives of the fallen vehemently and poignantly expressed their utter contempt for the protesters and, by implication, the protests – their words (as you might anticipate) were juxtaposed with familiar, yet ghostly, images of youthful soldiers going 'over the top' in the First World War. Instead of the dead burying the dead, the dead were burying the living under the cold earth of an unrealised future; the future that those men and women had died for. The ghosts altered reality: now, the protests were haunted by an idealised past *and* an idealised future. Shudders ran down everyone's spine. These ghosts are impossible to ignore, especially as they were being ill-treated in death, as they had been in life.

The imaginations of the living call to life the dead generations as they shimmer in photographs and old newsreels or in the testimonies of those who remember.[58] In these images and memories, a deathly question haunts British politics: who else had these men died for but 'us'? The debt the living owe these young, brave dead for their sacrifices, absolutely irredeemable. Certainly not by the protester(s) who had defiled the Cenotaph. And once the ghosts were on the side of authority and tradition, older political visions began to cast a dark shadow over the not-yet-born political visions of the protesters. The tradition of that dead generation weighed heavily on the attempts of the living to revolutionise traditions of democracy, freedom and justice. Suddenly, an alternative flow of history had been dammed up: the past stood firm, bitterly accusing the present of its failure to live up to its dreams for the future.

Ghosts take on many guises in the story above: we have large armies of men, in uniform, covered in blood; we have the spectres of dead ideas – revolutionary or not, militaristic or not, democratic or not; there is even the ghost of England's green and pleasant land; and of 'Jerusalem', the ideal city that London has so evidently failed to become. These spectres call forth ideas, and also feelings. The indignity and injustice of death returns (once again) to haunt the living. And the living are (once again) caught up in the traumas and losses of the past. The spectral geography of London now exposed and raw: haunted *simultaneously* by ghosts of the past, present and future. As Walter Benjamin might observe, in this one event we can see the ghostly means through which the dialectic of history – of the past, present and future – is brought to a stifling stand-still,[59] as the bony fingers of the dead tighten their deathly grip upon the living.

The question, however, is not whether we can simply remove the dead from life, but how best to live with the dead and their legacies (and this would include Marx and Marxism, of course). In his commentary on Derrida's *Spectres of Marx*, Fredric Jameson says this:

> **To forget the dead altogether is impious in ways that prepare their own retribution, but to remember the dead is neurotic and obsessive and merely feeds a sterile repetition. There is no 'proper' way of relating to the dead and the past. (1999, pages 58–59)**

Thus: to create an equitable, fair and democratic city, we must take into account the dead, but not become possessed by them. Perhaps this improper response to ghosts would appease Marx and Derrida. We cannot ignore the dead, otherwise we may never learn from them, nor will we honour them. But nor can we endlessly and melancholically return to the dead, lest we become unhealthily attached to them, lest we become entrapped in the relentless, drowning flow of their history. London's ghosts have, on occasion, proved to be a liability. What is true for London, is also true for New Orleans, Berlin, Singapore, Paris, New York and, possibly, every other city too.[60] Cities cannot simply give up the ghost. Even the physical structures themselves – or the gaps they leave behind when they 'pass on' – can become ghosts.[61] Listen to Virginia Woolf (but remember Walter Benjamin too):

> Where then can one go in London to find peace and the assurance that the dead sleep and are at rest? London, after all, is a city of tombs. But London nevertheless is a city in the full tide and race of human life. (1932, pages 125–126)

With so many opportunities for ghosts to suddenly (re)appear in the city, it is easy to see why the living might require constant reassurance that the dead rest easy. There remains an awkward accommodation between the living and the dead and future generations. The dead, as much as the living, must become part of a revolutionised history that is full of the flow of life. If we are to revolutionise the city, then we are faced with a series of dilemmas. Let me pick out two. We must free ourselves from the dead, but we must do so without simply denying the life of the dead: both their dreams and their nightmares. We must allow the dead to liberate us, but we must do so in ways that do not bind us to their long-gone dreams or, worse, their nightmares. Either way, the dead are a troublesome presence in the phantasmagoria of city life: where is the certain assurance that the dead rest peacefully and untroubled?

Ultimately, the just and free city will need to find ways to accommodate its ghosts (whether from the past, the present or the future; whether personal or collective; whether dead or alive) – as New Yorkers are all too aware, with intensely emotional conflicts over what to do with the World Trade Center site in the wake of 9/11 and the decision (in February 2003) to go with Daniel Libeskind's design for the rebuilding of the site.[62] Settling with ghosts may require a decisive, flexible and inclusive attitude to the physicality of the city, including its spatial form, its monuments and cemeteries, its sacred sites, its sites of trauma and injustice. Cities can seem such spirit-less places: modern, new, secular, all-too-real. In fact, they are far from (only) so, something rational planning and capitalist accumulation strategies often take too little account of. In this other-worldly spirit, perhaps, there are opportunities to commemorate and accommodate the dead anew.

In today's cosmopolitan London, perhaps the Cenotaph has temporarily outlived its usefulness. Though, maybe, it could be redesignated as a memorial to others who have died in armed conflict: perhaps commemorating those killed by the British in its imperial adventures; or, perhaps it might celebrate the victories of anti-colonial struggles. It is worth remembering, similarly, that among the 19,000 body parts supposedly to be stored at the World Trade Center site (awaiting identification) will be those belonging to the Al-Qaeda dead. 9/11 cannot be commemorated without remembering the 'anti-colonial struggle' it is forever tied to.[63]

Cities, to be free and just, must be inclusive and flexible in their treatment of their ghosts. They must be prepared to add, and take away, the dead. They must be able to see themselves in relation to old dreams and old hopes, but they must not let this after-life after-care stop them creating new dreams and new hopes. Cities must be carefully 'blasted' from the continuum of history in order to bring about alternative histories of the future. As the city contains many different pasts and futures, it may yet be possible to use one ghost against another to suggest alternative ways of proceeding. *That cities have so many ghosts might even prove an advantage.*

4.8 Conclusion: bury the dead, not the truth

In this chapter, I have told a number of ghost stories, each exemplifying something about the haunted and haunting emotional life of cities. Throughout, ghosts have set problems for the analysis of city life, not the least of which is the difficulty in getting them to say anything coherent. This is, then, my first observation. Ghosts are not coherent. They do not have one story to tell, or have one relationship to the living. There is no one loss, or trauma, or injustice, save perhaps death itself. Each ghost has its own unfinished business and it is rarely clear what this is. Ghosts stand between the business of life and the finality of death. They are threshold figures, existing between the worlds of the living and the dead, between appearance and disappearance, neither present nor absent, in a time and place of their own. Yet, this description can equally apply to the living. Ghosts easily join the phantasmagoria of city life because they are, in many respects, indistinguishable from the living. Perhaps for this reason, it is sometimes said that half the people you see in cities are ghosts.

The city is marked, then, by its multitude of ghosts; heterogeneous ghosts; a density of ghosts – the death mask of Wirth's definition of urbanism (1938). There is more to it, though. The city also haunts by commemorating its dead, in part by making them endure in its physical architecture. Thus, like other cities, London has its fair share of actual memorials (with more being added all the time[64]), but each new trauma creates a vernacular sacred architecture capable of calling forth the dead. Ghosts are part of the procession of figures through the phantasmagorias of city life. In this procession, the dead are juxtaposed, putting them in relation to one another – however silently – a spectral disruption of time and space caused by putting loss, trauma and injustice side by side. It is not simply headline events at landmark buildings that do this, for every now and then London roadsides (as elsewhere) sprout sad bunches of decaying flowers – so those who pass-by know someone has passed on. Cities haunt in the sense that they force us – perhaps against our will, perhaps only occasionally – to recognise the lives of those who have gone (before). In this sense, the physicality of the city itself shimmers with ghostliness as it becomes a mutable and durable place of memory.[65]

Though it may be opaque, ghosts have good reason to haunt the places they do. Ghosts often speak to some kind of trauma or injustice. Julie stands freezing on a New Orleans balcony to draw attention to her tragic fate, yet her presence also speaks to the injustice of slavery and the horrific psychodrama of colonial situations.[66] Ghosts also stand at the edges of the anxieties of the living. Ghosts appear in places where the provenance of a place is uncertain, or potentially unsettling or even dangerous. Examples of this can be seen in the ghosts that haunt modern blocks of flats. With a transient population, or with occult connections to the dead, it is impossible to know whether a flat might be haunted or not. Ghost stories therefore reveal *something* about the ways cities – such as Berlin, New Orleans,

New York, London and Singapore — accommodate their pasts, and how these pasts take hold of the present and the future. This *something* is a fractured emotional geography cut across by the shards of pain, loss, injustice and failure; an emotional world in which the ghost is the emblematic resident. More than this, the streets of New York and London — just as do the ghosts of other cities — take hold of the imagination: they defy us to find hope, since they have none; they defy us to make reparation, since none can be found for them; they defy us to allow them to bury themselves, since we have already buried them; they defy us to find new traditions, but in departing from the futures laid down by past generations, there is always the prospect of finally consigning them irrevocably to history.

I have suggested that ghosts manifest two different kinds of emotional state: first, they evoke the uncanny effects of city life; second, they allow us to see the grief-work of the city — its mourning, its melancholia. Tied to sites of haunting, ghosts can flash up at certain moments in cities to uncanny effect, reminding people of earlier losses and traumas of that place. For some, these are primarily personal anxieties and traumas. But seeing ghosts in cities allows the social anxieties and traumas to be witnessed also. Ghosts, then, do double duty here: they are simultaneously personal and social dramas. Similarly, haunting itself requires a double dialectical imagination: to see people as haunted (by the past), to see people as haunting (places); to see cities as haunted (by their pasts, by people, whether living or dead), to see cities haunting people. Ghosts and haunting, then, are spatial and temporal: located at a threshold that allows emotions and experiences to cross from one setting to another, in a flash, ghosts and haunting disrupt prosaic senses of time and place. Normally, such disruptions are guarded against: the smooth flow of history protected. Ghosts are the repressed returned: they shouldn't be there and there are consequences when they are.

Ghosts act as figures that condense many meanings, on to which can be displaced many feelings. Just like an element in a dream, the figure of the ghost is overdetermined — pointing in many different directions at once. This is no accident, perhaps.[67] In cities, the grief-work associated with ghosts is much the same as dream-work: emotions are shifted around dense networks of meaning and power. On the one hand, then, the difficulty with ghosts has been understanding what it is that they want. On the other hand, however, there is the dilemma of how to bury the dead without burying along with them what they have to say (Figure 4.16). The idea that

Figure 4.16
'Bury the Dead Not the Truth'.

cities might be engaged in grief-work emphasises the difficulty of living not only with the dead, but also with the truths they rarely ever fully embody. Cities have choices in how they deal with the dead. However, the lesson of grief-work is that decisions about commemoration and exorcism, remembrance and forgetting can never be made once and for all.

Ghosts disrupt notions of linear times and spaces: by returning, they alter both time and space, leaving neither quite where they had seemed to be. Ghosts, we can say, haunt the places where cities are out of joint – out of joint in terms of both time and space. They join the living world through fissures, sometimes these are called 'anachronisms', sometimes 'ruins': cities are full of them. They grip the imagination, screaming at the trauma and pain they have had to endure, lamenting the injustices that have befallen them. Such tales of pain and injustice are commonplace in cities. Often in sympathy, though perhaps trembling with a sense of the fragility of their own lives, people listen to the tales of loss and torment. Even so, this does not always lead to transformative recognition, nor even to the recognition of the social histories that lie behind ghost stories (especially where social dramas are converted, in ghost stories, into personal dramas). Ghosts can be a shocking presence in the phantasmagorias of city life, but not always shocking enough.

Conclusion
the real dream of city life

Introduction

The real city is characterised, as Virginia Woolf once put it, by 'the full tide and race of human life' – likewise, by perpetual movement, fervour, noise, buzz; by an incessant assault on the senses; by a ceaseless destructive creativity.[1] The effect of all this, according to Walter Benjamin, can be phantasmagoric. The city is a scene full of phantasmagorias. Things (of all kinds) pass through the lives of city dwellers, creating an immediate experience that looks for all the world like a procession of dream-like and ghost-like figures. In Benjamin's understanding, this experience resembles dreaming. The phantasmagoria is comprised of contradictory elements, which leave impressions: of a series of elements that can appear to be bizarrely put together, yet which city dwellers are usually indifferent to; of juxtaposed images and figures whose histories and geographies disrupt the smooth flow of time and space, yet in which history appears linear and singular and where geography seems to be merely a passive backdrop to the unfolding sequence of events; of a wide range of meanings and emotional intensities, yet which never quite match up with the urban scene – everything seems to have been produced elsewhere. In Benjamin's view, the swirl and whorl of the phantasmagorias can captivate the mind of the modern urban dweller, leaving them ... Hypnotised. Intoxicated. Stupefied. Dreaming. Sleeping.

In the big sleep of modernity, the city slumbers uneasily. Tossing and turning, as its nightmares and dreams take hold; the city is always threatening to wake up, but the disturbing dreams never quite jolt it awake. Modern folk seem to love their sleep: dream-images dance in front of their eyes, dazzling them, fascinating them. And why shouldn't they? The phantasmagorias of city life enchant and bewitch.[2] The immediate and necessary task is, for Benjamin, not only to look into these phantasmagorias, but also to figure out the work that goes into making them – to discover, that is, the phantasmatic means of production of urban space. For Benjamin, the real city is produced as much by ideas as machines, as much by fantasies as rational plans, as much by their psychodramas as their political economies.

Following Benjamin, as well as Simmel and Freud, there are many possible places to start the task of exploring the phantasmagorias of city life: the city is, after all, full of the tide and race of life, full of numerous phantasmagorias. Logically, though, it is best to begin with the idea of the phantasmagoria itself. If, as a visual experience, the phantasmagoria is defined by its dreaminess and ghostliness (literally 'made visible' in 'a place of assembly'), then an obvious starting point is with dreams and ghosts. Through empirical investigation, I have discovered that dreams and ghosts lead, by close association and cross-referencing, to magic and vampires. Indeed, such images and figures appear in the phantasmagorias of many cities, including New Orleans, New York, Berlin, London, Paris, Johannesburg and Singapore. Through these cities circulate the phantasmagorias of dreams, magic, vampires and ghosts; each city having its own unique pattern of, and relation to, these phantasmagorias. In this flow of images and events, it appears that more is going on than first meets the eye. In the phantasmagorias – as much as anywhere else – can be found the real life of the modern city.

Using a dream-analysis inspired by Freud, I believe it is possible to describe and interpret the work of phantasmagorias in the city. The idea of dream-work suggests a form of analysis that enables, on the one hand, the emotional qualities of city life to be evoked and also, on the other, the city's phantasmagoric elements to be tracked (spatially and temporally) through networks of affect, meaning and power. So, in the next two sections, I will explore, first, the emotion-work and, then, the space-work that goes into the production of the phantasmagorias of modern city life. As with dreams, it is often the marginal and seemingly trivial aspects of city life that yield the most telling information. And one more thing: for both Benjamin and Freud, in dreams there is redemption of a kind – for all dreams anticipate awakening. Dreams, they hoped, would provide the royal road out of the traumas and injustices of the past, allowing people to break free from the ghosts that haunt them in the present and which prevent new forms of life either from being envisaged or from emerging. There is a utopian imperative in this work which I will take up towards the end of this conclusion. Before we can get to this point, however, it is necessary to make explicit the different kinds of emotional work that are involved in the production of the phantasmagorias of city life.

The Work of Modern City Life

In *Real Cities*, encouraged by Freud and Benjamin, I utilise the idea of the dream and dreaming as a starting point both for exploring modern city life and for illuminating its 'unconscious' emotional processes, both personal and social. Employing the model of the dream for thinking about city life, I identify four elements of the phantasmagorias of city life for interpretation. These are dreams (themselves), magic, vampires and ghosts. From a Freudian perspective, these elements are assumed to be overdetermined; that is, determined by various meanings and affects many times over. Each element in a city's phantasmagoria is worked-over: meanings are condensed in each image or figure, while they are also sites through which, or on to which, feelings are displaced. In many ways, this accords with Benjamin's idea that phantasmagorias are also intensely technological: while they might appear superficial, they hide both the means of their production and also the labour, including the emotional labour, that goes into their production.

In this section, I draw out the emotional aspects of urban social processes that are almost always overlooked in analyses of real city life. I will explore the different kinds of emotional work that I have identified as shaping city life: namely dream-work, magic-work, blood-work and grief-work. Clearly, in any particular city, these processes are unlikely to be of equal importance (although, I guess, in some places they may be). Nor are they the only ones that can be identified using a dream interpretation of cities, nor that may be relevant to an understanding of the emotional life of particular cities – as I pointed out in the Introduction. Even so, in the cities I have explored, it is clear that dreaming, wish fulfilment and disappointment, anxieties and desires over blood and mortality, and a troubled relationship to the dead and death, are fundamental aspects of both the social processes and the emotional dynamics that produce city life and urban space. They are, moreover, significant in the various phantasmagorias of many other cities too. Now, let me develop each form of emotional work in turn, beginning with dream-work.

Freud used the expression **dream-work** to describe the work that goes into making dreams.[3] Put another way, Freud is arguing that a wide range of operations are performed on a host of elements, from memory and experience, to produce the fleeting experience known as 'the dream'. This work is highly complex and very clever: complex enough and clever enough to fool the dreamer, who usually remains asleep even in the most bizarre, or the most frightening, of circumstances. More than this, the dream creates scenarios in which feelings and meanings are disguised, in a whole variety of ways. The elements of any dream point in many directions at once, each suggesting other meanings. For Freud, each dream element is *overdetermined*: by meaning, by emotion, by sensation, by thought. In Benjamin's hands, this analysis of dreams is set within the social contexts of history, memory, place and power.[4] Dreams are both personal and social: indeed, they undermine any clear-cut distinction between these two realms. The city, in this view, is dream-like: each element in the urban landscape is overdetermined and there is a suppression, or disguising, of the desires and anxieties that underlie city life, including (or especially) those associated with capitalism and other relations of power.

Dreams are part of ordinary, banal life in cities. Every day, as we saw in section 1.2, people are confronted with dreams: dream homes with dream mortgages, dream loans for dream holidays, dream hair to go with the dream lingerie, and dream books to make sense of it all. While the rhetoric of dreams speaks most obviously to urban dwellers' yearning for a better life, cultural expressions of dreams are also overdetermined. In one direction, they can point to the wants generated by commodity culture; in another direction, they can suggest a general urban condition of unfulfilled desires – a lack of fulfilment that, arguably, commodity culture preys upon. And so on. Cities, in this respect, are part of the machinery of desire: they are sites where dreams are forged, housed, circulated. In dreams, the dreaming city reveals both its desires and its anxieties. This is serious. Dreams are not merely wistful fancies, but sites of the complex emotional work of real desires, personally and socially.

Neither Freud nor Benjamin made the mistake of seeing dreams outside their personal and social contexts. In seeing the city as made up through dream-work, it is possible to see the way in which personal and social imaginaries, personal and social desires and anxieties, are brought together and expressed – albeit in disguised, deceitful ways. Dream-analysis unearths the ways emotions circulate through cities: such as, the way that different moods and atmospheres become associated with different parts of the city; or how certain aspects of city life can seem unfeeling (often as an inversion of what people ought to be feeling); or the contradictory feelings that become attached to certain sites and cities. The emotional map of the city follows the twists and turns, the ups and downs, of affect as it travels through, and moves between, various aspects of city life.

It is tracking 'things',[5] along with the emotions associated with them, that proves invaluable in exploring the emotional qualities of city life. These emotional qualities are distinguished often by ambivalence and sometimes by irony. Thus, for example, dreams are often used to articulate innermost wishes, yet these wishes are rarely ever properly articulated and usually only disappointingly fulfilled. Nonetheless, dreams suggest a range of emotional mechanisms through which the phantasmagorias of city life are produced. These include condensations and displacements, networks of association, reversal into opposites, ambivalence, compromise and constant revision. And, occasionally, the uncanny return of the repressed.

While dream-work reveals how desires and anxieties gain (disguised) expression in the city through the widespread image of dreams, **magic-work** is oriented towards the practical and hopeful fulfilment of wishes in city life. Like dreams, magic is born of unfulfilled wishes and of unresolved anxieties. Unlike dreams, magic – to give emphasis to the 'doing' of magic – attempts to make wishes come true (or anxieties go away) by using its understanding of the occult world to intervene in the world, most famously using such magic-work as charms, rituals and even Voodoo dolls. Spells, for example, are produced from a variety of disparate elements: from a wish that wants to be fulfilled; to various magical elements, that are then deployed using an understanding of the real (occult) processes that underlie the world of appearances. Moreover, magic often involves rituals of production, such as incantations (prayers) or music (drums and dancing) or the use of particular devices (pentagrams, crosses, snakes), and specific sites that are either produced (e.g. by building a church or arranging things in a specific order) or have inherent magical properties (such as mountain tops or the intersection points of lay lines).

Through magic-work practices, certain beliefs in an occult world are animated. Occult relationships underlie how wishes and anxieties, beliefs and rituals, are combined in the attempt to make wishes come true. In the face of the bitter and banal disappointments of city life, magic offers other worlds through which wishes might come true: some appeal to deities to intervene on their behalf, while others seek to harness the energetics of the earth, and so on. In some accounts, even, the city itself seems to be a magical force, partly because it assembles and combines elements with diverse properties in particular, densely-woven, patterns.

Tracking magical beliefs, meanwhile, shows how ideas flow through, and interact within, cities. Magic – as in the case of Voodoo – shows that the city can convene a wide range of beliefs and syncretise them in highly specific ways. The city does not, however, melt occult relations into one coherent, integrated, rational urban belief system (which may be characterised as being civilised or scientific). Instead, differences can emerge out of the new rituals and spells that people invent to make the city work for them, for better or for worse. Magic-work, moreover, shows how people in cities attempt to fulfil their wishes through a variety of means. The ordinary labour of cities is often supplemented by a sense that city life might reap other rewards: perhaps through luck; perhaps through the intensification of the wish, often by greater concentration and continuous repetition (as in prayer).

The idea that magic-work goes into the 'doing' of city spaces and city life allows for different magical influences and purposes to be ascertained. Magic-work is not morally neutral. Nor is there consistency among and between magics. That magical beliefs and practices do not always have everyone's interests at heart is fundamental to the logic of magic-work. Magic-work, in other words, highlights a certain lack of compromise in city life – the extent to which self-interests are callously pursued – and it reminds us that some people get hurt in the process, as in the case of Adam (the ritually murdered young boy). The expression of a wish necessarily carries with it a health warning: be careful, be careful what you wish for, and, as likely, be careful of the wishes of others.

Finally, the phantasmagoria of magic leaves the impression that the city casts a spell over its citizens – for cities are believed to be places where people's wishes (for a better life, perhaps) can be fulfilled. Cities, themselves, seem to be the product of human wish-fulfilment: they are the magnificent achievement of the civilisation, Park insists. Despite and because of this, magical beliefs and practices are not expected to thrive in cities, but instead to fall upon fallow ground and die out. Highlighting the magic-work of cities also casts into sharp relief those beliefs and practices that are supposedly antagonistic to magic, especially science and secularism. Urban rationalities, nonetheless, turn out to be sympathetic, in many ways, to magic. Indeed, science, magic and religion are often co-existent, intimately-related, world-views and practices.

The city – as phantasmagoric – is never simply a place of fancy, visual pleasure or trickery. The city-as-phantasmagoria is also a place of bodies, of blood, of mortality. In many magical beliefs and rituals, thus, blood is an essential ingredient, speaking to both life and death. Life and death are embodied in blood, and with the smell of blood the figure of the vampire surfaces. Indeed, the affects associated with blood and the vampire are intimately connected. Through the social figure of the vampire, I throw daylight upon the vital **blood-work** of cities. As a city element, blood acts as a point of capture, or a focus of attention, for a wide range of emotions, meanings and powers. What the vampire shows is just how intense the overdetermination of the meaning of blood is within – and between – cities. Doing the blood-work of cities casts up a set of issues relating to some of the most fundamental elements of the emotional life of city: desires and anxieties over movement and belonging, horror and eroticism, mortality and immortality.

The vampiric city seeks to preserve its undead nature by draining the life-blood of its victims, an insatiable thirst that can never be quenched. Blood is at the heart of matters. Doing the blood-work of cities exposes the kinds of pleasure and horror bound up with the circulation, and spilling, of blood. Thus, many cities have regulations that govern such things as slaughterhouses and the handling of human bodies. Cleansing blood also highlights the effort undertaken by cities to marginalise, de-intensify, and localise the facts of death and mortality. The figure of the vampire enables us to see that cities do this to ensure that they continue to have a life after death. The city, in this light, becomes Un-Dead: the blood-work pointing less often to the life-giving social processes of the city, than to the blood exacted as the price of living in the city.[6]

The immortality, indiscernibility and mobility of the vampire raises questions about those features of cities that last longer than life, that circulate discontinuously, that arrive as unrecognisable and dangerous strangers, and that might want to extract a price in blood. Magic works by bringing elements of the city together, combining and arranging them in specific ways, and operating upon them, to produce wished-for effects. Vampires work rather by beguiling their victims, by changing form, vanishing from mortal time, reappearing after many a year, or simply by moving too quickly for the human eye. The vampiric city has a very different quality to the magical city. The vampiric city does not offer to fulfil any wish but that for life-after-death. Alongside the desire to be the undead vampire, though, there is also horror.

In cosmopolitan cities such as New Orleans and London, the unending flow of unnoticed lost lives – and, as Dracula argued, the ever-present existence of strangers and strangeness – is perfect for the vampire. And also for the vampiric city, as it feeds on diversity, on the life-blood of its populations, on the life-blood of an age at odds with itself; that is, on modernity. The vampire's appearances are covert, secretive, always under cover of the strange shadowy phantasmagorias of city life. Yet, modernity has its antidotes to the Un-Modern vampire. The vampire is always at risk of being moved on, chased out of the city. Under the sign of modernity are mustered a wide variety of technologies and techniques (such as those for recording life, from statistics to emails) as well as of homeopathic remedies, knowledges and institutions (from holy water to blood transfusions to international communities of doctors). Ranged together, these have often been successful in seeing off the vampire. Even so, the vampire always threatens to come back from the dead.

The Un-Dead vampire stands at the threshold between life and death. The vampire is not alone in this. The city, in some ways, is also between life and death, between a dead past and a vital future: splayed by the ever-fastening forces of progress and modernity. Vampire time – like city time – may be (almost) eternal, but it is not linear. The vampire does not live in a time that moves smoothly from beginning to end, the serial procession of life from cradle to grave. Time for the vampire jumps about, back and forth: sometimes on purpose, sometimes not. In this sense, the vampire is emblematic of the discontinuous circulations and heterotemporalities in the phantasmagorias of city life. The vampire reveals that the city itself comprises heterogeneous histories and geographies. The real danger, for vampires and cities, is that they lose contact with, or become exhausted by, their times and spaces.

The ghost, like the vampire, stands at the threshold of life and death – but the ghost haunts this threshold space differently. Like the vampire, the ghost speaks to the emotional forces of Life and Death. Unlike vampires, ghosts have a story to tell from beyond the grave. These stories may be difficult to understand, but ghosts' connection to memory, place and the past manifests an unfinished business – be it of loss, or tragedy, or revenge, or injustice, or whatever – that simply does not concern vampires. The ghost highlights most clearly the **grief-work** of city life. Grief-work is closely allied, as Freud observed, to dream-work. For this reason, it should not be much of a surprise that ghosts sometimes appear in dreams and that ghosts lend a dream-like quality to city life: indeed, make it *phantasmagoric*.

Evidence of grief-work can be found in every city, but this does not mean that all cities grieve in the same way (nor, of course, all people in any one city). Nonetheless, *all* cities have to deal with their dead: what grief-work points to is the significance of the choice of strategy for handling death, because there are consequences when the dead return to make demands upon the living. The dead can return, or hang around, in seemingly innocent ways – in memorials and commemorative rituals, in cemeteries and hospitals, in the names of buildings and streets, in personal acts of memory for those who have gone before. As many have observed, the city is a place of memory: of memories on top of memories, even of memories of memories. To repeat acts of remembrance for some events, while forgetting or exorcising others, is a mark of what – and who – haunts the city, and how. We can see this in individual ghost stories, such as Julie's fate in New

Orleans, as well as Singapore's Festival of the Hungry Ghosts and events after the defacing of the Cenotaph in London.

The ghost highlights the intense emotional work of grieving that is an essential part of the way cities deal with their dead, whether fairly or unfairly, in shame and guilt, in celebration or through forgetting. To understand this grief-work, I have drawn upon Freud's description of the uncanny and also his analysis of the processes of mourning and melancholia. These emotional registers suggest that the dead remain present in life, even when they are apparently absent. Wishes and fears associated with the dead hang around the living, haunting them, however coldly or occasionally. And the dead always threaten to return as ghosts, to make demands upon the living. Yet, cities are haunted by both people and ghosts — and their co-existence sets a series of moral problems: how does the city deal with the fundamental injustice of death? How are the living to co-exist with the aggrieved dead? Who, indeed, is to possess ghosts and what should ghosts be allowed to possess? When is it best to give up the ghost?

In many ways, the modern city seems antipathetic to ghosts. Living in the full tide and race of the new and the progressive, they should be indifferent to tragedies and traumas that ought to be dead and buried, consigned to a gone-by past. However, personal and collective senses of respect for the dead — or perhaps senses of shame and guilt — do not necessarily go away just because modern city folk have lives to lead. Indeed, these feelings towards the dead can even intensify over time. The recognition of the contributions of, or injustices that beset, previous generations offers a potential motivation for social transformation, much as Avery Gordon would wish. Even so, the persistence of a sense of injustice or of tragedy does not necessarily lead to transformative recognition and social transformation. Ghosts can beckon, they can haunt, but they do not always get their message across. The ghost may be a haunting figure in the phantasmagorias of city life, but this does not mean they always get paid proper attention.

The forms of emotional work that I have identified (dream-, magic-, blood-, and grief-) are rarely (if ever) taken into account when thinking about the production of city space and city life. Following Simmel, one way to accommodate these processes within an understanding of cities would be to build a psychological basement beneath urban political economy — with its circuits of capital, its labour processes, its commodity relations, its forms of consumption. Indeed, I think that Marx deliberately deployed a rhetoric of dreams, ghosts and vampires both as a political tool and also as a way of conveying the emotional qualities of political economy. Even so, the work of *Real Cities* has altered how urban economic processes should be viewed. (Re)Installing an emotional — psychological, even — appreciation of political economy means that it is now possible to appreciate the *ghost-like* dead labour in labour processes; the *vampire-like* immortal calculations and insatiability of capital accumulation; the *magical properties* of the commodity; and the *dream-like* quality of cultures of consumption. There are other forms of emotional work that can be brought under this (pirate) flag. I hinted as much in the Introduction to this book. What I am arguing is that forms of emotional work, such as dream-work, are not separate from the supposedly real work of the social processes of cities *but absolutely part of them*. In other words, the social processes that produce and reproduce cities and city life are inseparably bound up with forms of emotional work.

This raises the spectre of whether all cities are basically the same (as produced by a universal dream-work) or not. I have proposed that dream-work and grief-work are ways to understand the emotional qualities of city life and have shown how they are operative in producing cities. The implication is that the emotional quality of an individual city can be described and assessed, thus: City 1 has emotion A, while City 2 had emotion B; or City 3 had primarily emotion A with a bit of B, ... and so on. I risk being read as if I am arguing that there is a universal standard for measuring the emotional life of any particular city. On the contrary, I believe my analysis has shown the exact opposite: it has revealed the (production of) differences between cities, even while making clear the heterogeneous and often hidden links between them. It is at this point in the argument that 'space' becomes important: not only does it interfere with the idea that there might be a universal standard against which to judge or measure all cities, it also suggests a variety of ways to follow the emotional labour being conducted *within and between* cities. In the next section, I will explore the space-work that is part and parcel of city life.

Occult Spatialities and City Life: New York, New Orleans, London and Singapore

By using the term space-work, I intend to highlight the spatial means through which the appearance and effects of the various phantasmagorias of city life are produced and experienced. From dreams, we learnt that space-work is essential to the phantasmagorias of city life, not as a passive screen on to which the ghost-like or dream-like figures are projected, but as integral to the occluded production of the urban scene – as in the spatial production of the appearance and experience of the dream. Not only are various elements assembled in particular ways, but they are worked over many times *spatially*, predominantly by such mechanisms as condensation or displacement. In this light, *the city is overdetermined spatially*; determined by diverse spatialities many times over. Space-work, I have found, includes making networks and chains of association; displacement and condensation (the concentration of various 'things' on one space); procession, mobility and circulation (albeit non-linear in time and space); proximity, contiguity and juxtaposition; thresholds, presence and absence; haunting ('being there' includes inhabitation and possession in various forms); similarity, the mutability of form and shape; and so on.

Like dream-work in dreams, space-work privileges the ways that meanings and feelings are added or removed, de-intensified or intensified, as elements are shifted through networks of affect and meaning, both personal and social. In city life, these operations are never innocent of their constitutive power relations. Significantly, these rely on 'making space' in particular ways, for particular purposes, commonly disguising how they have done so. Thus, phantasmagorias appear to exist in apparently natural spatial forms. One such aspect of 'making space' is the production of nested spatial scales: an alchemy of territory and boundaries, from body and home through to the transnational and global. Another is the creation of commonplace spaces in which phantasmatic things are habitually housed: everything from railway stations to museums, exhibitions to shop windows. An appreciation of space-work, as developed initially in the analysis of dreams, has begged questions about how these seemingly natural city spaces came to be produced – as if asking how the projection, and screen, of a phantasmagoria works.

Dominant ways of looking at space, however, tend to regard only certain things as important or significant. Space-work, as with dream-analysis, contains an injunction to pay attention to those spaces, those fragments of city life, that are usually paid no mind. The interpretation of dreams leaves open the possibility that any trivial or seemingly obvious dream element may become highly significant through analysis. Indeed, it is as likely that a seemingly insignificant element in a dream will prove as important as an element that is obviously dramatic or bizarre. We can (also) look to the spaces of unremarked life to discover the social processes producing city spaces – and the spatial production of city life. In this light, it is perhaps because vampires and ghosts are so familiar – yet so strange – that they barely get a mention in urban analyses. But urban analyses have missed a trick, for vampires and ghosts occupy, and make, very different spaces in the city: as vampires bleed (across space), as ghosts haunt (space).

Exploring vampires, ghosts and the like, in the city, requires this dream-like, phantasmatic sense of the production of space: as much to do with the *occult spatialities* that make city spaces as to do with the visible world of buildings and infrastructures (even if they're below ground or high in the sky). Or, alternatively, the phantasmagoria of city life has as much to do with its production through occult spatialities as it does with visual and spatial technologies and experiences; with the procession of 'things' through the city; and with the dream-like, ghost-like quality of city life. These occult spatialities take on many forms and have a variety of consequences. Thus, magic revealed the intricate occult circulations between cities: from New York to Haiti, from New Orleans to West Africa, from London to Johannesburg. Voodoo, meanwhile, exposed the occult spatialities that (re)constituted the spaces both of Congo Square and of New Orleans.

An appreciation of the occult spatialities that produce urban phantasmagorias enables us to see the geographical under-labourings that create the seemingly unremarkable spatialities of city life. These spatialities are quite specific: they do not lead just anywhere, they are not universal, and they don't all intersect with one another. Indeed, and to be clear, space itself is not a universal, no more than is time. Here, I am not simply arguing that privileging the analysis of one form of spatiality, such as circulation or connectivity, will somehow fail to account for the heterogeneous productions of urban space. Instead, I wish to instigate a search for other kinds of spatiality (i.e. lived experiences of space) and productions of space, just in case they help us to see, or to wish for, different things from those already identified.

Globalisation, for example, is commonly thought of in terms of flows of people, goods, money and the like. It is based on a spatiality of connection: the world is becoming more and more connected, with a concentration of these connections on certain cities, primarily New York, London and Tokyo.[7] Vampires, however, give us a very different sense of what these connections might be like. Rather than the smooth and instantaneous connection posited in the so-called network society,[8] there is a sense of migration, flight, pursuit, evasion, dissimulation, rumour, wreckage, death, ancient money, and the need for good record-keeping. Most significantly, tracking the vampire gives us the sense of 'things' jumping in space and time: from Transylvania and Malaysia to Paris and London; from ancient to modern. Meanwhile, sometimes, the vampire's most effective weapon is *waiting* – waiting until knowledges about it have passed into myth, so that it is no longer considered to be a threat. The space-work of circulation, then, is like a fibrillating heart, erratically pushing the social life-blood through its veins, but not always when and where you might expect – or, indeed, want. Globalisation, from a vampire perspective, is a story about occult circulations beyond human senses of space and time, about how the world is bitten into and sucked dry.

In the already globalised network society, there appear to be no limits to the transfusion of people and capital from one place to another. Even if globalised cosmopolitan élites seem to move freely around the world – as Dracula does – such movements flow through specific arteries, from place to place. Vampire blood-work, moreover, discloses that these transfusions are not always successful or even desirable. Despite initially promising success, for example, the blood transfusions between Lucy Westerna and her gallant men ultimately failed. It simply is not the case that the space-work of transfusion – movements of people, information and capital from place to place – will lead to desired ends. Alongside this, there are many anxieties, hazards and horrors attached when such things flow unexpectedly free. The vampire-hunters show that the openness of cities to strangers from other worlds is not always welcome – and indeed that it might be catastrophic. And also that there are many, and increasing, preventative measures to stop fearful strangers (as figured by the vampire) from turning up unexpectedly.

The invocation of the term 'occult' implies that there are productions of spaces that are somehow secret or hidden or magical (in bad ways). In part, I do mean this. What happened to Adam (the murdered boy) has to be borne in mind. Occult connections between places do not necessarily produce the kind of cosmopolitanism that moderns might wish to celebrate. Voodoo, as I have described it, is emblematic of an intense social ambivalence towards new syncretic forms of cultural expression. To some it is a fearful, blasphemous cult. For others, it is a life-saver. Either way, Voodoo demonstrates how occult connections can produce new syncretic forms and to what purposes these can be put – and, in the case of Voodoo, this is so much more than stick dolls and dancing to drums. In this sense, Voodoo is spatially worked over: not just as a set of beliefs and practices into which meanings and feelings are condensed and displaced, but also through its occult connections, circulations and distributions between places – for example, through a Voodoo Atlantic that entwines New York and Haiti, Benin City and London. As an experience of movement and circulation, be it of beliefs in magic or of vampires travelling between cities, the phantasmagorias of city life are evidently full of the tide and race of life – as mysterious 'things' flow through cities, however discontinuously, as networks of interconnections criss-cross in and out of the city.

Occult geographies of circulation infuse all aspects of city life, but these spatialities cannot be considered as somehow lacking in history. Ghosts allowed chains of association to different pasts and temporalities to be tracked. Connected intimately to specific spaces, ghosts appeared on balconies, in apartment blocks, in graveyards. These phantasmagorias create a seemingly endless procession of heterogeneous 'things' through time-and-space, that merge into one another, that seem to have no depth – as in a dream. Taken together, the city becomes a composite image – a collage, perhaps – of personal and collective experiences and memories: where figures in the phantasmagoria constantly evoke associations with other places and other times. Ghosts, for example, expose the traumas and tragedies of the past in a place. *The city is a site where heterogeneous spatialities-and-temporalities co-exist*. Ghosts confirmed as much.

It is not simply the case that tracking figures, such as dreams and ghosts, through the phantasmagorias of city life has involved following them from place to place or from time to time. These figures have also revealed *threshold spaces* – sited at intersection points, between worlds – thresholds which have enabled other things, such as feelings and meanings, to cross over them. These thresholds are not simply open, but instead are porous.

Only certain things are permitted across the threshold. Moreover, like cross-roads, thresholds are as much points of departure as sites of meetings or passing-by.

Each phantasmagoria has its own version of meeting and of passing, of being between worlds. *Dracula*, for example, allowed passage between Transylvania and London, between the ancient and the modern, between life and death. Ghosts, between life and death, between disappearance and return, between materiality and immateriality, between presence and absence. In the threshold spaces of the city, moreover, new forms of subjectivity and sociation can *emerge*. Thus, Voodoo emerged in the syncretic spaces of Congo Square, in particular, and New Orleans, in general. Figures in urban phantasmagorias act as a bridging, or breach, point between different worlds. This gives us quite a different sense of the city. Cities, in this view, are 'out of joint' in time and space: stories from other contexts (always threaten to) disrupt its presence. Thus, ghosts bear echoes of the past and these reverberate into the future. Cities tremble as they are haunted.

Ghosts have affects and effects, differently, in different places in different times. Although ghosts, magic and vampires inform us about the emergence of new occult urbanisms, this does not mean that they can be treated as a gold standard against which the quality of life of a city — its secret desires and anxieties — can be assayed. As we have seen, New York is not London; Singapore is not New Orleans. Remember, for example, Vodou in New York and Voodoo in New Orleans; ghostly apparitions in Singapore and London; or the dream-worlds of Paris and Berlin. In the phantasmagorias of other cities' life, other phantasmatic elements may be more pertinent: say, for example, giants in Lund or angels in Budapest.

An analysis of the phantasmagorias of a city shows how dreams or ghosts or werewolves or angels are understood and the consequences of this for understanding the special qualities of city life there. For example, some people believe dreams are messages; for others, dreams are a way of dumping useless information. Whatever the view, there are consequences for actions taken in waking life, in the waking life of cities. It is the same with magic, vampires and ghosts. And angels, saints, dragons, werewolves, giants, clones and so on.[9]

It is space-work that selectively and carefully, perhaps deceitfully and cunningly, assembles these elements into the city's phantasmagorias: that is, space-work is part and parcel of the production of the phantasmagorias of any city's life. Importantly, *it is in among the phantasmagorias of city life that people seek ways to improve their lives, to conduct their projects, and to fulfil their wishes*. Each city must deal with its own form of space-work: the social and psychological processes that combine to make up its differentiated spaces.

In the analysis of the phantasmagorias of city life, dreams perform a triple duty: as an analytical model; as an object of study; and as the expression of a wish. Moreover, for Freud, dreams offer a royal road to people's hidden desires and anxieties. By following this royal road, Freud hoped that dreams would liberate people from — and also put them in contact with — their innermost desires. It is perhaps this that suggested to Benjamin that there might be something utopian about dreams. For Benjamin, the possibilities for dreams were most vivid at the point of awakening (somewhere between dreaming and waking life). For him, this was a revolutionary idea: Benjamin called for dreams to be taken seriously — dreams were something to learn from. More than this, moderns would have to learn to dream; dream while awakening; awaken to dreaming new histories. Dreaming itself was part of the utopian project.[10]

Dreaming New Futures for the City

Dreams, intuitively, seem to be a realm free from the constraints of social and cultural norms. Many people have used a dream-story to imagine what life might be like released from the dead-weight of tradition and history. Within the dream, plans for a perfect society can be laid out – seemingly unencumbered by the problems of actually bringing freedom and justice. Within these utopian dreams, the city has often proved to be the geographical setting for the perfect society, perhaps because cities appear to be the product of pure human endeavour. Likewise, Marxist geographer David Harvey has closed his eyes to dream his own utopian space-time. Harvey poses the vital question: what is to be done about the social problems in cities? Taking Baltimore as his example, Harvey outlines the nightmare of contemporary city life. His analysis is almost phantasmatic: an almost endless procession of images of decay, decline, death – yet, in the end, redemption might lie in a dream. (Or it might not.)

In *Spaces of Hope* (2000), David Harvey wishes to set out the resources of hope for thinking about a better, utopian, future. The title of the book, it should be noted, is itself a reference to Raymond Williams' last book, *Resources of Hope* (1989). And, thereby, is situated firmly in a long line of utopian thinking about cities, stretching back at least as far as Ebenezer Howard's *Garden Cities of Tomorrow* (1902).[11] However, Harvey refuses to allow himself to be constrained by debates either within urban planning or among contemporary urban critics. He also draws on a genre of imaginative utopian writings by novelists and science-fiction writers, linking his thought most explicitly with Edward Bellamy's utopian fantasy *Looking Backward, 2000–1887* (1887). Harvey's wish – perhaps like those of the May Day protesters (described in section 4.7) – is to revolutionise history, to dream anew the old dreams of utopia.

For Harvey, recovering utopian thought is a matter of urgency – the decline of utopian thought itself a symptom of the despair now pervading contemporary American cities (and presumably cities everywhere):

> Utopian longing has given way to unemployment, discrimination, despair, and alienation. Repressions and anger are now everywhere in evidence. There is no intellectual or aesthetic defence against them. Signs don't even matter in any fundamental sense any more. The city incarcerates the underprivileged and further marginalizes them in relation to broader society. (Harvey, 2000, page 11)

Harvey finds tangible evidence of the inequities and injustices of city life in his hometown of Baltimore: a city that he examined in his classic book *Social Justice and the City* (1973). Looking backward from 2000 to 1973, things could hardly be said to have got better: and they weren't exactly great in 1973! In Harvey's judgement 'Baltimore is, for the most part, a mess. Not the kind of enchanting mess that makes cities such interesting places to explore, but an awful mess' (2000, page 133). The extent of this mess is dispiriting: it is evidenced in vacant houses, homelessness, unemployment, the percentage of people receiving welfare payments, soup kitchens, charity missions, inequality, poor employment (involving the replacement of relatively well-paid, full-time industrial employment by low-paid, part-time and temporary service jobs), poor schools, low educational attainment, social distress, poor health, suburban sprawl (with its conformity and environmental unfriendliness), deindustrialisation, and a monstrous downtown.

In such circumstances, the spaces of hope are hard to find; for the most part, Hope seems to have died. Looking backward from 2000 to the 1880s, not much seems to have changed for the better. Reference can be made to Thomson's contemporaneous poem about the dreadful state of Victorian cities (1880). Part I of *The City of Dreadful Night* opens with the city as a dream that dissolves with *The* day, but – night after night – this nightmarish dream turns out to be real (page 29, lines 8–21):

> Dissolveth like a dream of night away;
> > though present in distempered gloom of thought
> And deadly weariness of heart all day
> > But when a dream night after night is brought
> Throughout a week, and such weeks few or many
> Recur each year for several years, can any
> > Discern that dream from real life in aught?
>
> For life is but a dream whose shapes return,
> > Some frequently, some seldom, some by night
> And some by day, some night and day: we learn,
> > The while all change and many vanish quite,
> In their recurrence with recurrent changes
> A certain seeming order; where this ranges
> > We count things real; such is memory's might.

For Thomson, the death of Hope, Charity and Faith had cast Victorian London into a perpetual night; a 'deadly weariness of heart all day'. Harvey is similarly concerned with the grim fate of Hope, Charity and Faith. More promisingly, for Harvey, all three sisters seem to be struggling on together in contemporary Baltimore: in its soup kitchens and its church missions; in its union meetings and its community activism. Thomson's melancholic hopefulness lay in the possibility that, in the real dream of city life, a new society might emerge. But a part of Harvey remains suspicious of such utopian day-dreaming. Where people are too despairing in the impoverished hand-to-mouth existence of the inner city or too taken with the glittering attractions of the soulless middle-class entrepreneurial city, utopian thinking appears to be a futile exercise in hopeless wishing. Despite the best efforts of Faith, Hope and Charity, then, Utopia seems to have died a death.

Yet, Harvey wishes to reanimate Utopia, as a guide to future action. To begin with, Harvey sets about a post-mortem of Utopia. This reveals two causes of death, each the (dialectical) opposite of the other.[12] On the one hand, utopianism has involved imagining a perfect place – this he calls *a utopianism of spatial form*: a city, for example, with a perfectly ordered spatial form. Such thinking can be seen in many places, but most recognisably in the urban plans of Ebenezer Howard and Le Corbusier. For Harvey, this is the triumph of spatial form over social process – a perfect society is stabilised by a perfect urban spatial order. Stick! No (more) change required. Such ideas, Harvey argues, have become inextricably bound up with authoritarianism and totalitarianism. And so creating a utopian map of spatial form has become a deathly exercise for anyone interested in freedom, justice and democracy.

On the other hand, Harvey identifies *a utopianism of social process*, which is associated with thinkers such as Marx and Hegel. Social processes, as they currently stand, will have to be revolutionised. However, it is impossible to determine in advance what kinds of spatial form might be appropriate or necessary for new revolutionised forms of social relation. For Harvey, since these revolutionary transformations remain endlessly open, it becomes impossible to intervene in the production of space as it already exists. In this formula for Utopia, social process determines spatial form. Utopia wastes away because, until the perfect revolution has taken place and its outcomes have been worked through, she has 'no place' to actually be.

Harvey's solution to this deadly dilemma is to suggest that a new kind of utopianism be explored: *spatiotemporal utopianism*. This utopia engages the utopianisms of spatial form and temporal process dialectically, to create the possibility of imagining the free and just city as both transformative of space-time and transformed in space-time. Visionary thought would be brought to bear on both social processes and spatial form at one and the same time. The intention, then, would be to empower people to create a utopian city built of transformative physical forms and socialised forms of governance and economy. Each stage in the freeing of the city – in its form and process – would necessitate further transformations in form and process. To get the ball rolling, Harvey suggests that empowerment might come through interventions in three areas: in government, in the economy, and in the creation of universal rights. Later in the analysis, Harvey identifies eight fronts on which political interventions can be made: transforming everything from the self to community; institutions to the built environment; ecologies to the universal principles of human action.

Inevitably, Harvey argues, such interventions carry with them the danger of authoritarianism and exclusion, but, without any interventions, the greater danger is that the catastrophe of city life will continue like this. *Now* is the time to begin to imagine the utopian city:

> There is a time and place in the ceaseless human endeavour to change the world, when alternative visions, no matter how fantastic, provide the grist for shaping powerful political forces for change. I believe we are precisely at such a moment. *Utopian dreams* in any case never entirely fade away. They are omnipresent as the hidden signifiers of our desires. (Harvey, 2000, pages 195, emphasis added)

He is serious about dreams of utopia – in an appendix to *Spaces of Hope*, Harvey uses the conceit of a dream to map out the fantastic possibility of an alternative future for cities. As with other utopian dreamers, his vision begins in the nightmare of the present-day. Eventually, Harvey dreams the dream of Utopia. Harvey's utopian dream begins in 2005, at a point when global warming begins to become undeniable. With growing inequalities of wealth and power, the global system comes under increasing strain. Somewhere between 2010 and 2013, a stock market crash, originating in Russia, stretches the world financial system to breaking point. The global capitalist system goes into melt-down, provoking a world-wide military take-over in 2014. The military are violent and oppressive and, to maintain their authority, install a highly advanced system of mass surveillance. Similarly, they create global standards of governance, communication, and the like. Nevertheless, by 2019, disparate and diverse oppositional movements somehow come together to create a global revolution.

The kind of oppositional politics Harvey is talking about can be found in recent protests around the world, such as those in London. For example, on 1 May 2003, a wide variety of different anti-capitalist groups began protests at various sites across London, especially

Figure C.1
Diverse protest groups gather opposite the offices of arms manufacturer Lockheed Martin, London, 1 May 2003.

those associated with the arms industry in the wake of the war against Iraq (Figure C.1). As Harvey might observe, these events hardly constitute a global revolution — but, in future, who knows? By the end of the day, police had herded many of the protest groups into Trafalgar Square. The police then began to don riot gear, just in case (Figure C. 2). As the sun went down, the riot police forcefully dispersed the good-natured crowd, using shields and batons to emphasise their (over) determination to prevent the violence that had marked previous May Day protests.[13]

In Harvey's dream, the global revolution is violently and ruthlessly suppressed — no surprise to those who witness the actions of militarised police forces around the world or indeed the military policing of the world (cruise missile diplomacy). For Harvey, though, at this point, working-class women create a global anti-military cultural movement of unstoppable force. This organisation begins in Buenos Aires and is known as 'The Mothers of Those Not Yet Born': it is non-violent and combines passive resistance with mass action.[14] It is, ultimately, successful. And (in part two of Harvey's dream) a utopian form of global spatial organisation is installed.

Figure C.2
Riot police near Trafalgar Square, London, 1 May 2003.

It is perhaps best to think of the utopian spatial form of Harvey's dream as a kind of globalised central place theory — this is somewhat in keeping, in fact, with the spirit of Christaller's own thinking, which had its utopian (or perhaps dystopian) elements![15] The nested hierarchies of spaces and places has, at its lowest level, 'hearths' (with 20–30 adults and kids) and/or 'pradashas' (parenting collectives); then 'neighbourhoods' (containing 10 hearths); 'edilias' (containing 200 plus neighbourhoods); 'regiona' (containing 20 to 50 edilia); 'nationa' (as a federation of regiona). It isn't that simple, though, since there are no fixed spatial scales and no fixed political organisations, especially since the dream system allows for the free flow of people and goods across *regiona*. Here, the 'dialectic of space and time' is most apparent.

However, the consequence of actually having no fixed scales, the free flow of people and goods and a situation of continual political upheaval is perhaps far too dream-like – far too chaotic and disorderly – for the logical playing out of space-time dialectics in Harvey's dream scheme. In the detail, we see that the scales are fixed, flows are controlled, exchanges between people regulated by a transparency enabled by those ever so useful military surveillance systems, and the political system of representative democracy is completely stable. In the end, the model appears to be 'utopic' because everything is collectivised: from sex and babies to goods and services. Even the psyche is collectivised: individuals get psychotherapy should their passions and desires get the better of them. Possession and obsession are, in the nicest possible way, *forbidden*.

On waking from his utopian reverie, Harvey warns that this dream is outlandish, outrageous and even nightmarish – perhaps because the dream seems so authoritarian, totalitarian; so lifeless. His 'waking self' is ambiguous – has he disowned it or not? Either way, Harvey's waking self remains interested in imagining a world without money, with free workers, without the ever-increasing pace of life, with respect and without competition (i.e. with collective collaboration). These are utopianisms of process, but for Harvey the waking lesson of the dream-world is that these ideals must happen *somewhere*: that is, that utopianism cannot remain simply a fanciful dream (or nightmare). Unless dreams lead to action, Baltimore will continue just as it is and the catastrophe of contemporary city life continues. The lesson for the future is clear: to produce the just and free city, something must be done now.

Harvey is desperate to wake up from the real nightmare of city life: the alarm bell rings loud. Wake Up! But let us remember Benjamin and Freud. And the analysis of the phantasmagorias of city life. Harvey's injunction is that the world must wake up *from* the dream-like – nightmarish – quality of modern life, into the open-eyed, brightly-lit dawn of a new utopian urbanism. Alternatively, following Freud and Benjamin, we might instead wake up *to* the hopes and wishes contained within the phantasmagorias of city life: with their emotional dynamism, their dreams and inspirations, their distractions and entertainments, their tears and laughter. Let us not be too hasty in abandoning dreams, in casting our most fanciful desires into the dustbin of history, while muttering bitterly about living in hell. Instead, critical inquiry into phantasmagorias is a good place to begin mapping out future forms of city life: with its suspicions about the appearance of things and about immediate experience; with its built-in emphasis on emotional work; with its recognition of the significance of space-work (and its collaborator, time-work). If we are to do something real about the nightmare of modern city life, then its phantasmagorias must be dealt with.

The Real of Cities and the Life of Phantasmagorias

To deal with the real city – the city as it really is – does not just mean taking into account its physical and social structures, but also its mental life. Simmel wished to build a basement below historical materialism that would house an account of intellectual culture; an account capable of explaining the valuations of, and the metaphysical and psychological preconditions to, economic life. I have been less concerned with constructing a basement below urban theory – and descriptions of city life – than with providing an alternative way of accounting for its infrastructures. However, like Simmel and Benjamin, I have been interested in the emotional life of cities and the phantasmagorias of city life. For me, this is what the real city is about. For an account of the affects and atmospheres of city life provides other ways of imagining cities as well as intervening in urban social processes. Thus, to dream the city anew, I have argued, following Benjamin, we must learn to dream. This is no less a real task facing urban theory than providing an accurate account of poverty or terrorism or housing or social exclusion or technological infrastructures or … or whatever the latest 'urban problem' is. To be clear, I am not privileging dream-analysis over other forms of analysis, but I am arguing that these other analyses lack something if they cannot account for, or engage with, the dreams and nightmares of city life.

I have not suggested, as Lefebvre has done before, that there is a real city, an imagined city or indeed a real-and-imagined city.[16] Instead I have argued that *the real city is overdetermined*: that there are multiple determinations acting upon city life and that these are always emotional in some way (even in their most rational guises). The names of only a few of these determinations are the physical, the social, the mental, fantasy, the imagined, memory, wishes, anxieties, affect, politics, sexuality, death, money. A critical account of real cities, then, must be capable of bringing these determinations into view, as and when necessary. Following the psychogeographers and the lessons of dream-analysis, I have begun by looking to socially marginal elements of city life, to show how vitally they connect to the real of city life: dreams as revolutionary theory, ghosts as political dilemma. In this spirit, I have tracked these elements, through various cities, in a variety of ways. I am hopeful that this has been worth it, for this analysis has both demonstrated the different forms of emotional work that go into making cities and also highlighted the co-existence and production of heterogeneous, phantasmagoric spaces and times. There is one further outcome to this research: it has shown just how real the phantasmagorias of city life are.

As I see it, a focus on phantasmagorias shows that real city life has a ghost-like, dream-like, emotional quality. In some ways, the idea that there are phantasmagorias to city life is grounded in a real 'smoke and mirrors' special effect, a trick of visual perception attuned to the procession and movement of innumerable things through the city. In one way, this phantasmagoric effect is produced by the sheer number of things going on in cities: the observer simply can't quite grasp what's going on, nor really determine where it is, or whether it has any solidity or consistency – any realness – about it. In another way, the movement of images and social figures through the city creates a phantasmatic psychodrama. This psychodrama is not always that dramatic – indeed, it is often characterised by a certain indifference or a blasé attitude (as Simmel would have it) – but it is nonetheless ordinarily experienced. Phantasmagorias act as a stage for these psychodramas, providing images and figures that condense and displace, personal and social, meanings and feelings.

Sometimes it can seem as if the urban dweller is like Klee's *Angelus Novus* (Figure C.3), only facing forwards, towards the future (not the past): watching the wreckage of progress flash past, unable to determine their place in history, unable to see either where things are coming from or where they end up.[17] The real catastrophe is that it continues like this, as Benjamin would say.[18] The furious storm of modernity still blows through city life. But city dwellers face *into* the wind, a contorted smile appearing on their faces – full of disappointment, yet ever hopeful of making it through the storm, heads held high. Even as the wreckage piles up behind their backs. Perhaps, there is hope in the image of a

Figure C.3
Angelic presence in New Orleans.

forward-facing *Angelus Novus*. The angel of the future might be able to dream up new times and spaces, perhaps by creating new forms of association, new forms of meaning and power. With such an angel at their side, city dwellers might not have to walk alone through the storm.[19]

The phantasmagoria of modernity is not simply made up of the illusion of forward progress to an inevitable future, nor is the wreckage of progress (in the modern city) only made up of the ceaseless obsolescence of new things. Just as much, the wreckage of modern city life is produced by its grey and colourless phantasmagorias, which are seen neither for what they are, nor for what they might offer. Phantasmatic images and figures slip through the fingers, like a dream upon waking. With hope, I have seized upon images and figures in the phantasmagorias of city life: upon dreams, magic, vampires and ghosts. Through them I have examined the emotional work and space-work that goes into the production of city life. None of these images and figures has pointed unambiguously towards the golden sky at the end of the storm of modernity. Even so, the work that went into making the phantasmagorias creates the possibility that city life might be made up, differently – if only through learning how to dream. In the phantasmagorias of city life, there are always possibilities for dreaming the real city anew.

Notes

Introduction

1   See also Pile, 1999.
2   See also Donald, 1999, pages 8–19.
3   See also Donald, 1997 and 2000.
4   See de Certeau, 1984; also, Sennett, 1970 and 1994; Soja, 2001; Maspero, 1990; Augé, 1986; and, historically, see Rendell, 2002.
5   Sinclair also uses other spatial practices to trace the mood of London: for example, in *Lud Heat* (1975), Sinclair draws lines (mapped by Marc Atkins) between Hawksmoor churches to suggest the hidden ancient, mysterious, mythical patterns underlying contemporary London life; in *Downriver* (1991), Sinclair tells stories strangely connected by the River Thames.
6   For a critical reading of Sinclair's *flânerie*, see Donald, 1999, pages 184–185; also, Pile, 2002. For a different reading of the neighbourhood, see Wright, 2001.
7   In some ways, Sinclair's injunction to 'notice everything' is similar to Freud's therapeutic practice of 'evenly suspended attention', enabling him both to notice everything and not to prejudge either their importance or history (see Laplanche and Pontalis, 1973, pages 43–45).
8   See Daniels, 1995.
9   See section 4.6.
10 See Debord 1955 and 1956; see also Blazwick et al., 1989; Bonnett, 1989; Plant, 1992; Pinder, 1996; Sadler, 1998.
11 On spatialities of power, see Allen, 1999b and also 2003.
12 See Debord, 1955, page 19.
13 An injunction that also applies to postcards, of course.
14 See Kern, 1983, page 15.
15 On the divisions – especially gendered – between public space and private space, see Imray and Middleton, 1983; Little, Peake and Richardson, 1988; and Spain, 2001.
16 See Robinson, forthcoming, Chapter 2.
17 On the history of manners and affect, see Elias, 1939.
18 Of course, the idea that the social processes producing city life are hidden has been a long-standing feature of urban analysis: some recent, and creative, versions of this can be seen in Sandercock, 1998; and, Borden, Kerr, Rendell and Pivaro, 2001.

19 See Buck-Morss, 1989, Chapter 8; Gilloch, 1996, Chapter 3. On Walter Benjamin's use of 'phantasmagoria', see also Britzolakis, 1999.
20 See also Berman, 1982, page 15.
21 The development of this analytical model began in another context: see Pile, 1998.
22 See Benjamin, 1927–40 [M1,6], page 418; similarly see Poe, 1839.
23 On the city and Jekyll and Hyde, see Donald, 1999, pages 117–119. For Donald, the 'monster within' is suggestive both of the splits within masculine sexuality and also of the uncanny that lurks within modernity.
24 See Schlör, 1998, and Schivelbusch, 1983, on the social histories of night and light in cities.
25 On Budapest's catacombs, see Kósa and Szablyár, 2002. On Statue Park, see Boros, 2002, and Nadkarni, 2003.
26 For a related argument, see Donald, 1999, pages 17–18, where he thinks about how angels might help us understand how the spectral spaces of the city are inhabited.

Chapter 1   The Dreaming city

1 It is no accident that 'dreams' should so often come into view in advertisements: 'Fashion, like architecture, inheres in the darkness of the lived moment, belongs to the dream consciousness of the collective. The latter awakes, for example, in advertising' (Benjamin, 1927–40 [K2a,4], page 393).
2 On the city as a dream factory, see Donald, 1999, pages 86–91.
3 It is not just in modern times that Augé discovers battles over dreams. According to Augé (following Le Goff, 1985), the Catholic Church in the medieval world waged a war on particular understandings of dreams. The Church had to make a strict division between legitimate and illegitimate interpretations of dreams. The basic problem, for them, was knowing whether a dream was a sign from God or the Devil, since dreams don't arrive signed and authenticated. That God-given signs may be delivered in dreams remains a widespread belief. Osama bin Laden, for example, is reputed to be a great believer in the prophetic quality of dreams (Saghiyeh, 2001, page 5). Dreams do not have to be messages from the unconscious self to the conscious self, nor messages from God or angels. Nor does every culture produce dream-books, but dreams are significant in many cultures (see, for example, Brody, 1981; and Tedlock, 1992; or, alternatively, Goh and Wong, 2001). For some, dreams can be something quite different, a world unto itself. Augé tells of African representational systems in which there is understood to be a continuity between waking and sleeping life (for another take on this, see Watts, 1999). For the Yoruba and the Ibo, for example, the dream is a journey through other worlds. The dreamer leaves the body to go on this journey, to return on waking. This explains why the body is so tired on waking: it is the weariness of the traveller. For me, these various interpretations of dreams are significant because they make a difference to city life.
4 Dreams, for Sobel, also expressed American desires and fears, especially in relation to race and gender.
5 It is best, therefore, only to buy one dream-book, otherwise you might never know the true meaning of your dreams!
6 This is to say that capitalism creates a dream-world of commodity fetishism, of course. However, specific ideologies, such as neo-liberalism, themselves dream of particular capitalist social relations: see Mitchell, 1999.

7   Helen McLean has published books on dream interpretation (including McLean and Cole, 2001) and runs her own website, www.catchadream.com.
8   There are several versions of this scenario, each with its own business nightmare.
9   See also Robinson, 1998.
10 It is not just politicians and activists who dream of freedom. In an episode of the popular 1970s British prison comedy, *Porridge*, two cellmates, Fletch and Godber, try to get through the night by imagining themselves in the world outside the prison. Since they cannot get out, they must create the world in their heads, as if in a dream. Fletch explains:

> FLETCH: Dreams is freedom.
> GODBER: Freedom?
> FLETCH: No locked doors, is there?
> (*Porridge*, episode 'A Night In')

11 Schnitzler's book provides the source material for Stanley Kubrick's last film, *Eyes Wide Shut* (1999).
12 See also Donald, 1999, pages 69–73, where he also discusses the role of dreams; or, alternatively, Vidler, 1992 and 2000; and Jacobs, 1996.
13 Gaiman has also written several novels, one of which, *Neverwhere* (1996), was turned into a BBC television series.
14 In this series, a young English boy called Timothy Hunter is given the option of choosing a life of magic or a life of science – a theme that will be taken up in Chapter 2.
15 I will return to John Constantine in section 4.4.
16 Morpheus's story parallels that of Laius, father of Oedipus, in some important respects. It might be useful to think about this in terms of Freud's suggestion of an Oedipal complex. However, I do not wish to engage Freud's mythological understanding of mental complexes in this book (instead, see Pile, 1996, Chapter 3).
17 Thus, for example, in the story 'Soft Places' (1992), Marco Polo is lost in the desert. In the intense heat, he is dreaming (probably in deliberate reference to Calvino's *Invisible Cities*). In his dream, Marco Polo meets G. K. Chesterton who states (page 141):

> Time at the edge of the dreaming is softer than elsewhere, and here in the soft places it loops and whorls on itself: in the soft places where the border between dreams and reality is eroded, or has not yet formed … Time. It's like throwing a stone into a pool. It casts ripples. Hoom. That's where we are. Here. In the soft places, where the geographies of dream intrude upon the real.

This is where we are in this chapter: in the soft places, where dream and reality loop, ripple and whorl. It is worth adding that G. K. Chesterton appears in this tale because he was a writer on dreams. Most relevant here is his short story, 'The Angry Street: a bad dream' (1908), which bears strong similarities to Gaiman's 'A Tale of Two Cities' (1993b).

18 Indeed, Robert is close to Death, cute elder sister of Morpheus.
19 See Freud, 1915.
20 See Freud, 1919; and Pile, 2000.
21 I will return to this idea most explicitly in section 4.3.
22 See, respectively, Boyer, 1983; and Sanders, 2002. Alternatively, Hélène Binet's photographs have been described as revealing 'the dream life of buildings' (Glancey, 2002, page 12).

23 Within psychoanalysis, see also Resnik, 1987; Segal, 1990; Flanders, 1993; States, 1997; Budd, 1999; Blass, 2002; Quinodoz, 2002. The broader impact of Freud's interpretation of dreams is explored in Bloom, 1987; Ferguson, 1996; Pile, 1998; and L. Marcus, 1999. The relationship between history and dreams is developed in a special double issue of the *History Workshop Journal*, which includes an interview with Hanna Segal (Pick and Roper, 2000). There is much more work that I could cite, but even this brief list should convey a sense of the continuing significance both of dreams and of Freud's work on dreams.

24 See Freud, 1905 for a demonstration of his use of a dream fragment in therapy; see Freud, 1907 on the use of dreams in literature; see Freud, 1908 for his analysis of the difference between a dream and a day-dream and why a dream is not a work of art; see Freud, 1917a, 1920, 1933 for later additions to his work on dreams. Other significant writings by Freud on dreams include 1901, 1911, 1916–17, 1923 and 1925a.

25 See Freud, 1913, pages 152–156.

26 Indeed, Freud's spatial imagination meant that he was able to use both the topographical (unconscious, preconscious and conscious) and the structural (id, ego, and super-ego) models of the mind without incongruity – and this is clear even in *The Interpretation of Dreams* (1900). It is as a result of this, perhaps, that Khan argues that the relationship between intrapsychic agencies in the mind creates a 'dream-space' (1974).

27 This idea appears throughout *The Interpretation of Dreams*, see especially pages 201–213 and 701–727.

28 See Freud, 1900, pages 330–331, 722 and 736.

29 See Freud, 1910, Third Lecture, pages 55–68.

30 See also Freud, 1926.

31 See also Freud, 1917b and 1919.

32 See Freud, 1900, pages 595–596.

33 See Freud, 1900, pages 607–608.

34 In particular, see Buck-Morss, 1989 and 2000; Gilloch, 1996; and Weigel, 1996.

35 In some ways, as a result of this analysis, Paris has become the capital of the literature on the nineteenth century: see, for example, Prendergast, 1996; and Sheringham, 1996.

36 See Tester, 1994; also Pile, 1996, Chapter 7.

37 As Benjamin does in his essays on Paris, completed in 1935, 1938, 1939b.

38 On the history of Trafalgar Square, see Mace, 1976.

39 I will return to this story in section 4.7.

40 See also Pile, 2000.

41 Although 'One-Way Street' is often seen as having a unique form, Benjamin's 'Ibizan Sequence' (1932b) also contains a series of 'analytical fragments', in some of which, similarly, he recounts his dreams.

42 Elsewhere, Benjamin suggests that each generation goes through stages, from sleep to awakening. In this sense, 'each epoch has [...] a side turned towards dreams' (1927–40 [K1,1], page 388).

43 See also Benjamin, 1935, pages 159 and 176.

44 See also Benjamin, 1927–40 [K2,5], page 392.

45 A term Benjamin takes from Proust (see 1939a, pages 111–112; and, for example, 1927–40 [K8a,1], page 403 and [N4,3] page 464).

46 See Benjamin, 1927–40 [K1,3], [N3a,3] and [N4,1].

47 Here we might think of works on cities as far apart in academic terms as Walkowitz, 1992, and Hannigan, 1998, respectively.

Chapter 2 The Magic City

1. That is, the household gods.
2. This probably connects to Benjamin's argument about 'dialectics at a standstill' (Benjamin, 1927–40 [N3,1], page 463; and 1940). The threshold would be a space where dialectics are caught. In this case, magic might describe what emerges from (a reanimated) dialectics.
3. Sigmund Freud makes a similar argument in his own work on civilisation (1921, 1930): see also Gilman, 1993. Echoes of this are also to be found in Geography: Robert Sack, for example, describes myth and magic as being far from those of (as he describes it) objective western beliefs (1980, Chapter 6).
4. Such an idea also underpins beliefs in astrology: that the distribution of far-off planets can influence the personality and destiny of the individual. On astrology, see Adorno, 1974a. Adorno's argument that magic is a form of regression is also pursued in 1974b.
5. See also Taussig, 1997; and Stephen, 1998. This also connects to Freud's arguments about the relationship between the return of the repressed and fear of the dead: see Freud, 1939, pp. 372–376. See also Chapter 4 of this book.
6. Here, Park's arguments are close to Simmel's: see Introduction.
7. Park does not cite Freud, but 'borrowing' is suggested both by the strong similarities in the argument and also by the shared use of the example of rain-making in elaborating the difference between magic and science (Freud, 1913, page 138; Park, 1925a, page 127): in magic, individuals mimic the rain; in religion, they ask the gods for rain; in science, they seed the clouds.
8. It is significant that Freud sees the history of civilisation corresponding to the history of individuals. This allows him to correlate 'primitives' with so-called savages and half-savages with children, women and importantly neurotics (see Pile, 1996, Chapter 4).
9. Alternatively, see Bloom, 1999.
10. It is interesting to note that many bookshops in the USA, the UK and Singapore now sell DIY Voodoo doll kits!
11. Similarly, telepathy: see Freud, 1913, page 142.
12. There is continued support for such a view: see, for example, Stephen, 1999.
13. That is, less than a decade before Park published his paper, and only about a quarter of a century before the last witch trial in Britain.
14. Here, Penczak is clearly arguing against certain back-to-nature forms of witchcraft. There are, of course, a wide variety of witchcrafts, but even modern forms of witchcraft, such as wicca, have a tendency to make little acknowledgement of, or concessions to, modernity (whether urban or not): for example, see Farrar and Farrar, 1981, especially Chapter XI.
15. Penczak's spelling of magic with a k is possibly a reference to the writings of Aleister Crowley (e.g. 1929).
16. There are echoes here of debates about whether the city is natural or not: for different views, see, for example, Tuan, 1978; Cronon, 1991; or Harvey, 1996.
17. The British Museum is associated with magic in other ways. The Museum houses some of the magical devices owned and used by Dr John Dee (1527–1608/9). Moreover, although the bulk of his library of 'natural philosophy' (or magic) was destroyed or scattered, some of Dr Dee's books ended up – via Elias Ashmole and Hans Sloane – comprising one of the three major collections of the British Museum's library when it was founded in 1753 (see Harkness, 1999, page 220). For this reason, Tim Brennan claims

that 'the British Museum is founded on magick!' (2003, page 2). On Dee's life and times, see French, 1972; Clulee, 1988; Harkness, 1999; and Woolley, 2002; see also Cosgrove, 1990; and Harrison, Pile and Thrift, 2004, pages 12–42. It is worth noting the close links between magic (calculing) and mathematics (calculating) in Elizabethan times, as this gives a new twist to the idea that a defining characteristic of modern urban mentalities is their calculativity (after Simmel).

18 Most writings on Voodoo are about Haitian Vodou: see, for example, Deren, 1953; Métraux, 1959; Davis, 1985; and Fleurant, 1996. On the circulation of Voodoo between Haiti and the USA – a Vodou economics – see also Browning, 1998, Chapter 5.

19 Indeed, Vodou has been a long-standing part of Haitian nationhood. The slave rebellions of 1791 (see James, 1938), led by Toussaint L'Ouverture among others, made use of Vodou to protect themselves against the bullets of the slave-owners and their armies. Indeed, Vodou rites provided the medium for rebellion (1938, page 86), in part led by a High Priest, Boukman. Insurrections were also given spiritual guidance by a Vodou priestess, Romaine. Further, the slaves consecrated their loyalty to each other and their willingness to fight to the death through the Vodou rite, Bois-Caïman. Vodou physician Francios 'Papa Doc' Duvalier seized power in a military coup in 1956 and established a dictatorship in 1964 ruthlessly supported by a Vodou-named militia, the Tonton Macoute. On Papa Doc's death in 1971, his son 'Baby Doc' took over and remained in power until 1986. Though a Catholic and an ex-Priest, when Jean-Bertrand Aristide became President in April 2003, he ensured that Vodou was adopted as an official religion of Haiti. A much rehearsed joke goes that Haiti is 70% Catholic, 30% Protestant and 100% Vodou. In February 2004, Aristide was forced to abandon Haiti by armed rebels. Vodou rites were an essential part of the rebellion in the north of the country.

20 There is a large literature on Voodoo in New Orleans, much of it written for a tourist market (though none the less fascinating for that). On the more academic side, there is Tallant, 1946; and Brolin, 1990. The less academic side has mainly been interested in Marie Laveau and her tomb: see Klein, 1996. An analysis that situates Marie Laveau within a women's history of New Orleans can be found in Gehman and Ries, 1988. For an attempt to draw Voodoo into European traditions of magic, see Black and Hyatt, 1995.

21 Congo Square, in fact, has been the crucible in which many cultural forms have been syncretised. Specifically, Congo Square has been seen as the birthplace of jazz and associated forms of music and dance: see Johnson, 1991; also Roach, 1996, Chapter 2.

22 See, for example, Hurbon, 1993.

23 See James, 1938.

24 See also Trouillot, 1995; and Slater, 2003, page 426.

25 See also Woods, 1998; or alternatively, Neate, 2001.

26 As you might expect, the history and meaning of Marie Laveau and her life is contested: see Fandrich, 1996.

27 A good sense of the significance of this can be gained from Taussig, 1997.

28 St John's Day is 24 June and All Saints Day is 1 November: we will hear more about Halloween in New Orleans below.

29 One of these tours, the Haunted History Tour, offers its own book: see Smith, 1998.

30 Various graphic novels have also drawn on New Orleans' Voodoo connections: for example, Ennis, 1997–98; and Moore, 1999. To ensure the webs are cross-referenced, it is worth noting that Penczak cites Morrison's *The Invisibles* series – one tale from this series begins in New Orleans (Morrison, 1998–99), and it even makes a passing reference to Freud (page 47).

31 That is, myself and two companions, Dydia DeLyser and Helen Regis.
32 It is for this reason that the Singapore one dollar coin has an eight-sided design, since this is a 'magical' symbol associated with *feng shui*. US dollars, similarly, have magical devices, including a (broken) pyramid and an all-seeing eye, possibly derived from freemasonry (similarly, see Gilbert, 1998). On economies as the circulation of signs, see Lash and Urry, 1994. On the sociology of money, see Dodd, 1994; and McDowell, 1997. On space and money, see Corbridge, Martin and Thrift, 1994; Leyshon and Thrift, 1997; and Allen and Pryke, 1999.
33 This is a reference to the cosmology of J. K. Rowling's *Harry Potter* books.
34 For example, in May 2002, a man was arrested in Krugersdorp, Johannesburg, for killing a 52-year-old man and cutting off his head in a 'muti murder'. Indeed, many South African police forces have special units devoted to witch-related crimes. See Niehaus, 2001; also Ashforth, 1996; and Niehaus with Mohlala and Shokane, 2001.
35 The role of the *sangoma* (or *izangoma*) in traditional medicine in southern Africa is divination. They decide whether the illness of a person is primarily bodily or spiritual. If it is bodily, the patient goes to an *izinyanga*, or herbalist, for a muti cure. Spiritual illnesses, such as possessions, require other cures, such as exorcism. It is worth noting, in the context of the discussion of Voodoo above, that these beliefs and practices have little to do with Voodoo, which is primarily West African.
36 On Vodhun (Voodoo) in West Africa, see Lovell, 2002.
37 On globalisation and witchcraft, see also Comaroff and Comaroff, 1993; and Geschiere, 1998.

Chapter 3 The Vampiric City

1 For an analysis that resonates with the one presented here, see Browning, 1998.
2 Blood is, indeed, life. In June 2001, a gang of 25 drug addicts were arrested for selling blood illegally to blood banks and hospitals in Delhi and Meerut, India.
3 Following Freud, 1926.
4 This is about the inability to recognise an absent object. The archetype for this formulation is the phallus.
5 See Zizek, 1989.
6 In August 2000, a man threw himself from the first-floor window of a house in Fitzrovia, London. The man believed, psychiatrists told the inquest, he was being pursued by vampires.
7 *Epigram*, in November 2001, reported that a former Bristol University secretary could be Bristol's first self-confessed vampire. The ex-secretary, who had renamed herself LaCroix, admitted to drinking human blood, from willing donors, and to sleeping in a velvet-lined coffin. It was not easy being a vampire, she revealed: 'There are things you just can't do. Going on holiday, going to the beach and doing the simplest of things like the shopping is a nightmare' (cited on page 5).
8 These are not the only recent vampire killers. On 24 November 2001, 17-year-old Matthew Hardman stabbed to death 90-year-old Mabel Leyshon in her home, in Anglesey, north Wales. Afterwards, he cut out her heart and put it in a saucepan. He placed pokers in a cross and a candlestick next to her body. Then he drank her blood. This ritual, he believed, would make him into an immortal vampire. As part of the evidence at the trial, police showed computer records of visits Hardman had made to

Figure 3.16 Queen of the Damned adverts appeared on London buses that passed the Chapel where the Queen Mother was lying in the state, April 2002.

vampire, and vampire/donor, websites. On the day Hardman went to trial, the papers reported that three men and a woman had been arrested in the Ukraine for a similar crime (for both stories, see Getty, 2002, page 7). In August 2002, Hardman was convicted and sentenced to a minimum of 12 years. Hardman lost an appeal against his conviction in January 2003.

On 11 December 2002, 22-year-old Allan Menzies killed 21-year-old Thomas McKendrick in Fauldhouse, West Lothian, Scotland. Menzies told the courts that he was under instructions from the female vampire Akasha, a character in the Anne Rice novel (and film of the same name) *Queen of the Damned* (Figure 3.16). After killing McKendrick, Menzies drank his blood and ate his flesh. During his trial, he declared he was now immortal and a vampire. In October 2003, Allan Menzies was convicted and sentenced to a minimum of 14 years.

9 Sometimes, they even organise day-trips, for example to the open-days for London cemeteries.
10 On fearful nights in London, see Schlör, 1998.
11 An image repeated in the film *Underworld* (2003). The (human) Romeo and (vampire) Juliet narrative of *Underworld* involves a 'class' war between, on the one hand, aristocratic and decadent (and in this way sexy) vampires and, on the other, workmanlike and honest (and to this extent sexy) werewolves. It is worth observing that vampires are both male and female, but it is hard to spot the female werewolves, and certainly no female werewolf is depicted in the midst of transformation. Significantly, the film is set

in Budapest, which doubles for the strangeness of East European cities, so close in the imagination to Transylvania.

12 On this idea that London was (is?) the heart of Empire, see Jacobs, 1996.
13 See Rickels, 1999, page 15; see also, Frayling, 1991, page 69.
14 On anaemia and vampirism, see Trigo, 1999. On the relationship between medical practices and vampirism, see White, 1995; and Haraway, 1995. Haraway (1995) also draws out a critical relationship between blood, kinship and science.
15 See also Baudelaire, 1857c.
16 See Benjamin, 1938, 1939a.
17 See Marx, 1850, page 88; 1852, page 242; 1867, page 342.
18 See also Dolar, 1991.
19 And other media besides. As you might expect, the comic-book hero Batman has encountered vampires (Delano, 1995), including the original Bat-Man, Dracula (Moench, 1991).
20 Perhaps ironically, this castle (Bran Castle, Transylvania, founded in 1377) is now a site for international tourism. For a description and history of Bran Castle, see Praoveanu, 1999.
21 On the detective and the city, see Donald, 1999; and Frisby, 2001.
22 See Moretti, 1978b.
23 It is for this reason, I think, that Alan Moore makes Mina Murray a key character in his steampunk comic series, *The League of Extraordinary Gentlemen* (1999–2000), set in an 'alternative' London of 1898. Intriguingly, it is implied that Mina Murray is herself a vampire (though she is explicitly so in the movie version). As one of the characters, Campion Bond, says, 'The British Empire has always encountered difficulty in distinguishing between its heroes and its monsters' (page 6). On monsters and cities, see Ruddick, 2004.
24 On Dracula (both person and book) as an allegory of the circulation of capital, see Moretti, 1978a; and Gelder, 1994, pages 17–20. For Gelder, the image of the vampire is directly correlated with the image of the Jew in both Marx and Stoker (pages 13–16). In this sense, Dracula also shows us the colour of money as well as the race of the vampire.
25 As he does in the movie *Dracula 2000* (2000), when travelling between London and New Orleans. Bram Stoker, however, would have known that there are no direct flights between the two cities and he would have devised a more realistic travel plan.
26 For a near contemporary account, see Engels, 1845.
27 See Young, 1995.
28 See Stallybrass and White, 1986, Chapter 3; Osborne, 1996; Kaïka and Swyngedouw, 2000.
29 See Corbin, 1986; Classen, Howes and Synnott, 1994.
30 Indeed, there's a close relation between the (self) regulation of the city, the (self) regulation of bodies and the regulation of self: see Elias, 1939; Foucault, 1963, 1984; Gay, 1984.
31 Corbin, 1986, Chapter 6.
32 The cleansing of blood from the streets of the city is, arguably, a case study in the abject: see Douglas, 1966; and Kristeva, 1980. For the geographies of abjection, see Sibley, 1995; and Robinson, 2000.
33 One of the Louisiana lottery scratch cards is titled 'Creepy Cash'. It is adorned by Frankenstein, a house like the one in *Psycho*, a scythe carrying Death *and* a vampire bat.
34 See also Schopp, 1997; and Williamson, 2001.
35 See, Choon, 2004. Related tales in the region concern 'man-hungry' widow ghosts: see Bell, 1995.
36 See, for example, *True Ghost Stories*, 2000a and 2000b; or Moey, 1990.

37 One interpretation of the sexiness of red lips is that they resemble the lips of the vagina. One more possibly relevant – and also possibly untrue – anecdote: in Roman times, prostitutes would rouge their lips to indicate their willingness to give oral sex.
38 See Moretti, 1978a, pages 100–101.
39 See, for example, Jackson, 1981; Carter, 1988.
40 See also section 2.4.
41 There are parallels to be drawn from the spatial treatment of slaves and the dead, both being given specially designated sites, side-by-side, *outside* the city limits. See section 2.4.
42 Florence, 1996, page 5.
43 See Roach, 1996.
44 See also section 4.1.
45 See also Bartolini, 2003.
46 Claudia is a young female vampire made by Louis. Their perverse incestuous and paedophilic relationship lies at the core of the novel. In both novel and film, the uncanny presence of sexuality in the figure of the child is fully played out. Sexuality is also at stake in Rice's novels. Vampires are queer (see Gelder, 1994, pages 58–64; see also Dyer, 1988), raising the spectre of perverse desires – but then usually killing them off.
47 See section 1.7. See also Caillois, 1937.
48 See Corbin, 1995.
49 As in Poe's famous purloined letter (1844).
50 Indeed, the term vampires was often used in the nineteenth century to refer to criminals. For example, the *Bodie Evening News*, around about 1878, referred to the 'bums and vampires' that populated the Californian mining town. (My thanks to Dydia DeLyser for drawing my attention to this.) Similarly, while responding to the threat Dracula poses to Gotham City, Batman observes: 'Let's just say … to me, all criminals are 'vampires' cloaked in night to prey on the innocent' (Moench, 1991, page 22).
51 The difference between human and vampire blood is a common theme of vampire tales, see for example the first *Blade* film (1998). The hero Blade is both a mix of human and vampire blood, and also a black vampiric figure (played by Wesley Snipes).
52 Another entertainment is actually going on these tours: I'd certainly recommend them!
53 A relevant work is Browning, 1998: Browning also raises issues such as Voodoo, Voodoo economics and, what I have been calling, a Voodoo Atlantic.

Chapter 4 The Ghostly City

1 Much has been made of ghosts being in a time 'out of joint'. This largely stems from Derrida's discussion of Hamlet: see, for example, Derrida, 1993, pages xix, 17 and *passim*; see also Laclau, 1995; and Jameson, 1999.
2 See also Bell, 1997. On trauma and history, see Hodgkin and Radstone, 2003.
3 See also Lee, 1995a, pages 46–64.
4 See National Archives of Singapore, 1996, page 48.
5 See National Archives of Singapore, 1996, pages 67–70.
6 The stories about massacres on Siloso Beach are unverified, but it is certain that bodies from massacres were washed up on Blakang Mati (Sentosa Island) beaches.
7 See Choon, 2004; Faucher, 2004.

8   On the Festival of the Hungry Ghost, see Goh, 1997; or, alternatively, Lee, 1995b; and Ong, 2001. Buddhists also celebrate the Festival of the Hungry Ghosts, but it honours filial loyalty, based on the legend of Mu Lian. For examples of encounters with hungry ghosts, see Stephen Fong's account of meeting his brother, Jimmy (2001); a clerk's story (Anon., 1989); or a housewife's tale (Anon., 2002).
9   See Freud, 1913, pages 114–116: see also Stephen, 1998 and Barrows, 1999.
10 For many, the festival is associated with auctions for 'lucky' goods, often ordinary household items such as wine or microwaves or piggy banks, which are sold at inflated prices, with profits often going to charity.
11 On this, see also Castle, 1988. See also S. Marcus, 1999, page 117.
12 See Gilloch, 1996; also Pile, 2002; on the place of memory, see Nora, 1989.
13 See also Hetherington, 2001; and Gilloch, 1996, Chapters 2 and 3.
14 See also Pile and Thrift, 2000; and also Pile, 2002.
15 See sections 1.4 and 1.5.
16 Contemporary Berlin is still trying to resolve its relationships to the dead: see Till, 2005. It is worth noting, in this context, that St Louis Cemetery No. 1 has a policy of segregating the dead. There is a plot laid aside specifically for Protestants (i.e. the damned!).
17 For example, of his memories of going to the theatre, Benjamin says 'in the end I can no longer even distinguish dream from reality' (page 334).
18 I doubt it is a coincidence that Freud also dreamt of ghosts who blocked his path: see his 'non vixit' dream (1900, initially pages 548–553).
19 On the uncanny, see Royle, 2003. On the uncanny in architectural practice, see Vidler, 1992. See also Castle, 1995. In different, but related, contexts, see Jacobs, 1996, Chapter 5; Cohen, 2000; and Gelder and Jacobs, 1998.
20 C. Sear: get it?
21 The stranger, of course, is an intensely ambiguous figure – someone both to rely on and to be afraid of: see Jacobs, 1961; Young, 1990; Wilson, 1991; and also Pile, Brook and Mooney, 1999.
22 On violence and city life, see Sandercock, 2002.
23 See section 1.6.
24 Burial grounds are popularly thought to be inhabited by ghosts. The catacombs of Edinburgh, for example, are widely believed to be haunted. This prompted psychologist Dr Richard Wiseman to conduct a 'scientific' study of the phenomena (Scott, 2001; also Rankin, 2001). During the ten-day research period, volunteers reported a range of ghostly happenings, from apparitions to physical contact. About half the people in certain vaults reported ghostly activity, while only a third in others. What the ghost-rich vaults had in common was that they were colder, darker and smaller. Dr Wiseman suggested that people felt more anxious in these kinds of space and were therefore more likely to 'experience' ghostly manifestations. Despite seeming to offer a scientific, psychological explanation for ghosts, the research actually lends itself to an opposite conclusion – that ghosts actually exist: it is just that they prefer to haunt small, dark, cold places! This study has, paradoxically, made tours of the catacombs even more popular. Not only are ghosts popular, and common, it's very hard to prove they're not there.
25 Constantine was first mentioned in section 1.3.
26 See Gehman and Ries, 1988, pages 12–15.
27 See Young's *Colonial Desire* (1995), especially Chapter 6 on white power and desire around miscegenation.

28 A story in the tradition of the 'tragic mulatto': see Doane, 1991, Chapter 11.
29 Similar stories are told in Rio de Janeiro, see Jaguaribe, 2001.
30 An alternative version can be found in Smith, 1998, pages 22–26. In this version, the master promises to marry Julie, but only if she spends an entire night naked on the balcony. Not for a moment does he think she will do this. One night, the master goes downstairs to drink and play chess with friends. Later, in the small hours, he eventually retires to bed. But Julie is nowhere to be seen. In a panic, he rushes to the balcony only to find her dead. Here, blame is shifted from the master to the mistress. More than this, Julie's temperament shifts from being loyal and quiescent to being wilful and tempestuous. In both, her love makes her at least foolhardy, at worst stupid.
31 Avery Gordon (1997, Chapter 4) makes a similar point in relation to Toni Morrison's novel, *Beloved* (1987). Similarly, see Ware and Back, 2002, Chapter 7. See also Sharpe, 2003.
32 See Brooks, 1982.
33 See also hooks, 2003.
34 Lee himself relates other tales of haunting from the same tower block, one involving a person who awoke to find a ghost tidying up the flat. For another haunted tower block tale, see Chan, 2001.
35 See also Mighall, 1999, Chapter 3.
36 In this light, there is a certain gothic quality to Benjamin's writings: for example, 1925–26. On ruins, see also Edensor, 2001.
37 See Kennedy, 2003. For another way of reading the bones of London, see Moorcock, 2001. By tracking bone, Moorcock makes spatial links between London, New York and Cairo, and temporal links between pasts, presents and futures. Or, alternatively, Garner, 2003.
38 See also section 3.4.
39 This book was turned into a film, *Bringing out the Dead* (1999), by Martin Scorsese.
40 See, for example, 1925b.
41 When Park suggested that the city is 'a state of mind' (1925a, page 1), he probably wasn't inferring that it was haunted by its pasts in the same way that people are.
42 See Freud, 1913, pages 92–93 and 107–122; see Freud, 1917b, for a development of these themes.
43 For an account of this eye-catching building, see Powell, 1992; or, alternatively, Pile, 2001.
44 MI is short for Military Intelligence.
45 A popular BBC TV spy drama series dealing with MI5 is titled *Spooks*. Intriguingly, the exterior views of Thames House (the name of MI5's main building in London) are not of Thames House, but of the HQ of another secret organisation, the Freemasons.
46 They were really *ear* witnesses, for almost no-one had actually seen anything!
47 See also the Introduction.
48 MI5 (security service) is listed in the telephone directory (with an address at 12 Millbank, just over the river from MI6), but there is no listing for MI6. The telephone directory once listed the address for the Government Communications Bureau (an alleged pseudonym for MI6) as being 85 Albert Embankment. Now, the Bureau only has a P.O. box number. Even so, the Bureau does have a land-line number. The exchange number, 7582, is Vauxhall's. My guess is the spooks still haunt 85 Albert Embankment.
49 For Degen (2001), this 'performance' of place is about the relationship between the body, the senses and spectral appearances.
50 See Introduction.
51 On Trafalgar Square as a site of protest, see Pile, 2003.

52 According to some newspaper reports: see, for example, *The Guardian*'s front page, 2 May 2000.
53 See St Clair, 1999.
54 Some speculated that as many as 30,000 officers were present or on call in central London that day.
55 Cited in the *Guardian*, 2 May 2000, front page.
56 As Lefebvre noted, 1974.
57 See, for example, Johnson, 1994, 1995.
58 On photography and ghosts, see Durant, 2003; Kaplan, 2003; and Schoonover, 2003.
59 See Benjamin, 1935, and also 1940.
60 On New Orleans, see Roach, 1996. On Berlin, see section 4.2; Ladd, 1997; and Till, 1999.
61 See Kuftinec, 1998; and DeLyser, 1999 and 2001.
62 That Libeskind also designed the Jewish Museum in Berlin is surely no accident: a disconcerting link has thereby been made between 9/11 and the Jewish holocaust. See also, Sorkin, 2003.
63 This anti-colonial struggle seems to be a special mix of anti-western, anti-American, anti-Christian and anti-capitalist geopolitical logics.
64 See also Kerr, 2001.
65 See also Curtis, 2001.
66 See also Gelder and Jacobs, 1999.
67 For Freud, melancholia and dreams are much alike (1917b, page 251), since they're both intimately connected to narcissistic mental disorders: on narcissism, see Freud, 1914.

Conclusion

1 See Woolf, 1932, page 126. A fuller citation can be found in section 4.7.
2 See also Bennett, 2001. Some would argue that modern life is not enchanting enough: see Moore, 1997.
3 See sections 1.4 and 1.5.
4 See sections 1.6, 1.7 and 4.2.
5 That is, by back-tracking through chains of association.
6 An analysis that might extend in many directions, including, for example, blood donation: see Titmuss, 1997.
7 See Sassen, 2001; and Taylor, 2003. For a counter argument, see Robinson, 2002 and forthcoming.
8 Here I am thinking of Castells, 1996, 1997a, 1997b.
9 Some of the 'social figures' that can be tracked are religious: for an interesting approach to thinking through such figures and city life, especially in terms of redemption, see Ward, 2000.
10 The idea that people dream in rational plans for cities is already well established: see Boyer, 1983. Examples would include Le Corbusier's designs for radiant cities: see Le Corbusier, 1927 and 1933. His designs for cities of light and movement also exposed his desire for an orderly city, a cleansed white city (and also his anxieties that cities might be otherwise): see Wilson, 1998. People's dreams, however, often turn out to be more chaotic, more devious, more unruly than rational plans take into consideration.
11 For a review of such schemes to emancipate cities, see Lees, 2004.
12 For a related diagnosis, see Mooney, Pile and Brook, 1999.

13 See Introduction and section 4.7.
14 There are ghosts here: for example, Sarah Radcliffe (1993) has talked about the mothers of the disappeared in Argentina and their intractable haunting of political resistance to the military dictatorship.
15 See Barnes, 2004.
16 See Lefebvre, 1974; and also Soja, 1996.
17 To play with Benjamin's description of the angel of history (1940, page 249). It is quoted on page 139.
18 'The concept of progress must be grounded in the idea of catastrophe. That things are 'status quo' *is* the catastrophe' (Benjamin, 1927–40 [N9a, 1], page 473).
19 YNWA.

References

Adorno, T. (1974a) 'The Stars Down to Earth: the *Los Angeles Times* astrology column' in T. Adorno, *The Stars Down to Earth and Other Essays on the Irrational in Culture*. 1994, London: Routledge, pp. 34–127.

Adorno, T. (1974b) 'Theses against Occultism' in T. Adorno, *The Stars Down to Earth and Other Essays on the Irrational in Culture*. 1994, London: Routledge, pp. 128–134.

Allen, J. (1999a) 'Worlds in the City' in D. Massey, J. Allen and S. Pile (eds), *City Worlds*. London: Routledge/The Open University, pp. 53–97.

Allen, J. (1999b) 'Cities of Power and Influence: settled formations' in J. Allen, D. Massey and M. Pryke (eds), *Unsettling Cities: Movement/Settlement*. London: Routledge/The Open University, pp. 181–218.

Allen, J. (2003) *Lost Geographies of Power*. Oxford: Basil Blackwell.

Allen, J. and Pryke, M. (1999) 'Money Cultures after Georg Simmel: mobility, movement, and identity', *Environment and Planning D: Society and Space*, **17**, pp. 51–68.

Anon. (1989) 'Lost Soul' in R. Lee (ed.), *True Singapore Ghost Stories, Volume 1*. Singapore: Angsana Books, pp. 21–23.

Anon. (2000a) 'Lone Woman' in *True Ghost Stories. Vampires Nightmares*. Singapore: ASURAS, pp. 30–34.

Anon. (2000b) 'Midnight Pontianak' in *True Ghost Stories. Vampires Nightmares*. Singapore: ASURAS, pp. 119–124.

Anon. (2002) 'Hungry Ghosts' in R. Lee (ed.), *All New True Singapore Ghost Stories, Volume 3*. Singapore: Native Communications, pp. 3–10.

Ashforth, A. (1996) 'Of Secrecy and the Commonplace: witchcraft and power in Soweto', *Social Research*, **63(4)**, pp. 1183–1234.

Ashforth, A. (2001) 'On Living in a World with Witches: everyday epistemology and spiritual insecurity in a modern African city (Soweto)' in H. Moore and T. Sanders (eds), *Magical Interpretations, Material Realities: modernity, witchcraft and the occult on postcolonial Africa*. London: Routledge, pp. 206–205.

Augé, M. (1986) *In The Metro*. 2002, Minneapolis: University of Minnesota Press.

Augé, M. (1997) *The War of Dreams: exercises in ethno-fiction*. 1999, London: Pluto Press.

Barnes, T. (2004) 'Central Places' in S. Harrison, S. Pile and N. Thrift (eds), *Patterned Ground: the entanglements of nature and culture*. London: Reaktion Press, pp. 179–181.

Barrows, K. (1999) 'Ghosts in the Swamp: some aspects of splitting and their relationship to parental losses', *International Journal of Psychoanalysis*, **80**, pp. 549–561.

Bartolini, N. (2003) 'A City in the Savage Garden: New Orleans in the eyes of Ann Rice's vampires'. Paper presented at the annual conference of the Association of American Geographers, New Orleans. Copy available from author.

Baudelaire, C. (1857a) 'The Metamorphoses of the Vampire' in C. Baudelaire, *The Flowers of Evil*. 1993, Oxford: Oxford University Press, pp. 253 and 255.

Baudelaire, C. (1857b) 'To A Woman Passing By' in C. Baudelaire, *The Flowers of Evil*. 1993, Oxford: Oxford University Press, p. 188.

Baudelaire, C. (1857c) 'The Vampire' in C. Baudelaire, *The Flowers of Evil*. 1993, Oxford: Oxford University Press, p. 65.

Bell, M. B. (1995) 'Attack of the Widow Ghosts: gender, death, and modernity in Northeast Thailand' in A. Ong and M. G. Peletz (eds), *Bewitching Women, Pious Men: gender and body politics in Southeast Asia*. Berkeley: University of California Press, pp. 244–273.

Bell, M. E. (2001) *Food for the Dead: on the trail of New England's vampires*. New York: Carroll and Graf Publishers.

Bell, M. M. (1997) 'The Ghosts of Place', *Theory and Society*, **26**, pp. 813–836.

Bellamy, E. (1887) *Looking Backward, 2000–1887*. Harmondsworth: Penguin.

Benjamin, W. (1925–26) 'One-Way Street' in W. Benjamin, *One Way Street and Other Writings*. 1979, London: Verso, pp. 45–104.

Benjamin, W. (1927) 'Dream Kitsch: gloss on surrealism' in W. Benjamin, *Selected Writings, Volume 2: 1927–1934*. 1999, Cambridge, MA: Belknap Press/Harvard University Press, pp. 3–5.

Benjamin, W. (1927–40) *The Arcades Project*. Cambridge, MA: Harvard University Press.

Benjamin, W. (1932a) 'A Berlin Chronicle' in W. Benjamin, *One Way Street and Other Writings*. 1979, London: Verso, pp. 293–346.

Benjamin, W. (1932b) 'Ibizan Sequence: Ibiza, april-may 1932' in W. Benjamin, *Selected Writings, Volume 2: 1927–1934*. 1999, Cambridge, MA: Belknap Press/Harvard University Press, pp. 587–594.

Benjamin, W. (1935) 'Paris – the capital of the nineteenth century. <Exposé of 1935>' in W. Benjamin, *Charles Baudelaire: a lyric poet in the era of high capitalism*. 1973, London: Verso, pp. 155–176. Alternative translation in W. Benjamin, *The Arcades Project*. 1999, Cambridge, MA: Harvard University Press, pp. 4–13.

Benjamin, W. (1938) 'The Paris of the Second Empire in Baudelaire' in W. Benjamin, *Charles Baudelaire: a lyric poet in the era of high capitalism*. 1973, London: Verso, pp. 9–106.

Benjamin, W. (1939a) 'Some Motifs in Baudelaire' in W. Benjamin, *Charles Baudelaire: a lyric poet in the era of high capitalism*. 1973, London: Verso, pp. 107–154.

Benjamin, W. (1939b) 'Paris – the capital of the nineteenth century. Exposé <of 1939>' in W. Benjamin, *The Arcades Project*. 1999, Cambridge, MA: Harvard University Press, pp. 14–26.

Benjamin, W. (1940) 'Theses on the Philosophy of History' in W. Benjamin, *Illuminations*. 1973, London: Fontana, pp. 245–255.

Bennett, J. (2001) *The Enchantment of Modern Life: attachments, crossings, and ethics*. Princeton, NJ: Princeton University Press.

Berman, M. (1982) *All That is Solid Melts into Air: the experience of modernity*. London: Verso.

Bhabha, H. (1985) 'Signs Taken for Wonders: questions of ambivalence and authority under a tree outside Delhi, May 1817' in H. Bhabha, *The Location of Culture*. 1994, London: Routledge, pp. 102–122.

Black, S. J. and Hyatt, C. S. (1995) *Urban Voodoo: a beginner's guide to Afro-Caribbean magic*. Tempe, AZ: New Falcon Publications.

Blass, R. (2002) *The Meaning of the Dream in Psychoanalysis*. New York: State University of New York.

Blazwick, I., in conjunction with Francis, M., Wollen, P. and Imrie, M. (eds) (1989) *An Endless Adventure ... An Endless Passion ... An Endless Banquet. A Situationist Scrapbook*. London: Verso.

Bloom, C. (1999) 'Angels in the Architecture: the economy of the supernatural' in P. Buse and A. Stott (eds), *Ghosts: deconstruction, psychoanalysis, history*. Basingstoke: Macmillan, pp. 226–243.

Bloom, H. (ed.) (1987) *Sigmund Freud's* The Interpretation of Dreams. New York: Chelsea House Publishers.

Bonnett, A. (1989) 'Situationism, Geography and Poststructuralism', *Environment and Planning D: Society and Space*, **7(2)**, pp. 131–147.

Borden, I., Kerr, J., Rendell, J. and Pivaro, A. (eds) (2001) *The Unknown City: contesting architecture and social space*. Cambridge, MA: The MIT Press.

Boros, G. (2002) *Statue Park*. Budapest: City Hall.

Boyer, C. (1983) *Dreaming the Rational City: the myth of American city planning*. Cambridge, MA: The MIT Press.

Brennan, T. (2003) *Museum of Angels: guide to the winged creatures in the collection*. London: British Museum/Gli Ori.

Brite, P. (1992) *Lost Souls*. Harmondsworth: Penguin.

Britzolakis, C. (1999) 'Phantasmagoria: Walter Benjamin and the poetics of urban modernism' in P. Buse and A. Stott (eds), *Ghosts: deconstruction, psychoanalysis, history*. Basingstoke: Macmillan, pp. 72–91.

Brody, H. (1981) *Maps and Dreams: Indians and the British Columbia frontier*. Vancouver: Douglas and McIntyre.

Brolin, R. (1990) *Voodoo: past and present*. Lafayette, LA: University of Southwestern Louisiana.

Brooks, J. (1982) *Ghosts of London*. Norwich: Jarrold Publishing.

Brown, K. M. (2001) *Mama Lola: a vodou priestess in Brooklyn* (Updated Edition) 1991, Berkeley: University of California Press.

Browning, B. (1998) *Infectious Rhythm: metaphors of contagion and the spread of African culture*. New York: Routledge.

Buck-Morss, S. (1989) *The Dialectics of Seeing: Walter Benjamin and the Arcades Project*. Cambridge, MA: The MIT Press.

Buck-Morss, S. (2000) *Dreamworld and Catastrophe: the passing of mass utopia in East and West*. Cambridge, MA: The MIT Press.

Budd, S. (1999) 'The Shark Behind the Sofa: the psychoanalytic theory of dreams', *History Workshop Journal*, **48**, pp. 133–150.

Caillois, R. (1937) 'Paris, a Modern Myth' in C. Frank (ed.), *The Edge of Surrealism: a Roger Caillois reader*. 2003, Durham, NC, and London: Duke University Press, pp. 176–189.

Calmet, A. (1746) 'Treatise on the Vampires of Hungary and Surrounding Regions', extracts in C. Frayling (ed.), *Vampyres: Lord Byron to Count Dracula*. London: Faber & Faber, pp. 92–103.

Calvino, I. (1972) *Invisible Cities*. London: Faber & Faber.

Carayol, R. and Firth, D. (2001) *Corporate Voodoo: principles for business mavericks and magicians*. Oxford: Capstone Publishing Company.

Carter, M. L. (1988) *Dracula: the vampire and its critics*. Ann Arbor, MI: UMI Research Press.

Castells, M. (1996) *The Rise of the Network Society: The Information Age: economy, society and culture Volume 1*. Oxford: Basil Blackwell.

Castells, M. (1997a) *The Power of Identity: The Information Age: economy, society and culture Volume 2*. Oxford: Basil Blackwell.

Castells, M. (1997b) *End of Millenium: The Information Age: economy, society and culture Volume 3*. Oxford: Basil Blackwell.

Castle, T. (1988) 'Phantasmagoria: spectral technologies and the metaphorics of modern reverie', *Critical Inquiry*, **15(1)**, pp. 26–61.

Castle, T. (1995) *The Female Thermometer: eighteenth-century culture and the invention of the uncanny*. Oxford: Oxford University Press.

Chan, E. F. L. (2001) 'Our Block of Flats is Haunted' in *There are Ghosts Everywhere in Singapore, Volume 3*. Singapore: The Publishing Consultant, pp. 34–37.

Chesterton, G. K. (1908) 'The Angry Street: a bad dream' in G. K. Chesterton, *Daylight and Nightmare: uncollected stories and fables by G. K. Chesterton*. 1986, London: Xanadu Publications, pp. 61–65.

Choon, B. K. (2004) 'Ghosts, Spectres and the other Presence' in R. Bishop, J. Phillips and W.-W. Yeo (eds), *Beyond Description: Singapore, space, historicity*. London: Routledge.
Classen, C., Howes, D. and Synnott, A. (1994) *Aroma: the cultural history of smell*. London: Routledge.
Clulee, N. (1988) *John Dee's Natural Philosophy: between science and religion*. London: Routledge.
Cohen, P. (2000) 'From the Other Side of the Tracks: dual cities, third space, and the urban uncanny in contemporary discourses of "race" and class' in G. Bridge and S. Watson (eds), *A Companion to the City*. Oxford: Basil Blackwell, pp. 316–330.
Comaroff, J. and Comaroff, J. (eds) (1993) *Modernity and its Malcontents: ritual and power in postcolonial Africa*. Chicago: University of Chicago Press.
Connelly, J. (1998) *Bringing out the Dead*. London: Warner Books.
Copjec, J. (1994) *Read My Desire: Lacan against the historicists*. Cambridge, MA: The MIT Press.
Corbin, A. (1986) *The Foul and the Fragrant: odour and the French social imagination*. 1994, London: Picador.
Corbin, A. (1995) 'The Blood of Paris' in *Time, Desire and Horror: towards a history of the sense*. Cambridge: Polity Press.
Corbridge, S., Martin, R. and Thrift, N. (eds) (1994) *Money, Power and Space*. Oxford: Basil Blackwell.
Cosgrove, D. (1990) 'Environmental Thought and Action: pre-modern and post-modern', *Transactions of the Institute of British Geographers*, **15(3)**, pp. 344–358.
Crary, J. (1990) *Techniques of the Observer: on vision and modernity in the nineteenth century*. Cambridge, MA: The MIT Press.
Cronon, W. (1991) *Nature's Metropolis: Chicago and the Great West*. New York: W. W. Norton and Company.
Crowley, A. (1929) *Magick in Theory and Practice*. 1976, New York: Dover.
Curtis, B. (2001) 'That Place Where: some thoughts on memory and the city' in I. Borden, J. Kerr, J. Rendell and A. Pivaro (eds), *The Unknown City: contesting architecture and social space*. Cambridge, MA: The MIT Press, pp. 54–67.
Daniels, S. (1995) 'Paris Envy: Patrick Keiller's *London*', *History Workshop Journal*, **40**, pp. 220–222.
Davis, W. (1985) *The Serpent and the Rainbow*. New York: Simon & Schuster.
de Certeau, M. (1984) *The Practice of Everyday Life*. London: University of California Press.
Debord, G. (1955) 'Introduction to a Critique of Urban Geography' in L. Andreotti and X. Costa (eds) (1996), *Theory of the Dérive and Other Situationist Writings on the City*. Barcelona: Museu d'Art Contemporani de Barcelona and ACTAR, pp. 18–21.
Debord, G. (1956) 'Theory of the Dérive' in L. Andreotti and X. Costa (eds) (1996) *Theory of the Dérive and Other Situationist Writings on the City*. Barcelona: Museu d'Art Contemporani de Barcelona and ACTAR, pp. 22–27.
Degen, M. (2001) 'Sensed Appearances: sensing the "performance of place"', *Space and Culture*, special issue on Spatial Hauntings, **11/12**, pp. 52–69.
Delano, J. (1995) *Batman: Manbat*. 1997, three parts in one volume, New York: DC Comics.
DeLyser, D. (1999) 'Authenticity on the ground: engaging the past in a California ghost town', *Annals of the Association of American Geographers*, **89(4)**, pp. 602–632.
DeLyser, D. (2001) 'When Less is More: absence and landscape in a California ghost town' in P. Adams, S. Hoelscher and K. Till (eds), *Textures of Place: exploring humanist geographies*. Minneapolis: University of Minnesota Press, pp. 24–40.
Deren, M. (1953) *Divine Horsemen: the living Gods of Haiti*. London: Thames and Hudson.
Derrida, J. (1993) *Spectres of Marx: the state of the debt, the work of mourning, and the new international*. 1994, London: Routledge.
Dickens, C. (1848) *Dombey and Son*. Harmondsworth, Penguin.
Doane, M. A. (1991) 'Dark Continents: epistemologies of racial and sexual difference in psychoanalysis and the cinema' in *Femmes Fatales: feminism, film theory, psychoanalysis*. London: Routledge, pp. 209–248.

Dodd, N. (1994) *The Sociology of Money: economics, reason and contemporary society*. Oxford: Polity Press.

Dolar, M. (1991) '"I Shall Be With you on Your Wedding Night": Lacan and the uncanny', *October*, **58**, pp. 5–23.

Donald, J. (1997) 'This, Here, Now: imagining the modern city' in S. Westwood and J. Williams (eds), *Imagining Cities: scripts, signs, memory*. London: Routledge, pp. 181–201.

Donald, J. (1999) *Imagining the Modern City*. London: Athlone Press.

Donald, J. (2000) 'The Immaterial City: representation, imagination, and media technologies' in G. Bridge and S. Watson (eds), *A Companion to the City*. Oxford: Basil Blackwell, pp. 26–34.

Douglas, M. (1966) *Purity and Danger: an analysis of the concepts of pollution and taboo*. London: Routledge.

Durant, M. (2003) 'The Blur of the Otherworldly', *Art Journal*, **62(3)**, pp. 6–17.

Dyer, R. (1988) 'Children of the Night: vampirism as homosexuality, homosexuality as vampirism' in S. Radstone (ed.), *Sweet Dreams: sexuality, gender and popular fiction*. London: Lawrence & Wishart.

Edensor, T. (2001) 'Haunting in the Ruins: matter and immateriality', *Space and Culture*, special issue on Spatial Hauntings, **11/12**, pp. 42–51.

Elias, N. (1939) *The Civilizing Process: sociogenetic and psychogenetic investigations*. 1994, Oxford: Basil Blackwell.

Ellis, B. (1993) 'The Highgate Cemetery Vampire Hunt: the Anglo-American connection in Satanic cult lore', *Folklore*, **104**, pp. 13–39.

Ellis, W. (1999) *Hellblazer: haunted*. 2003, parts 134–139 in one volume, New York: DC Comics.

Engels, F. (1845) *The Condition of the Working Class in England*. 1987, Harmondsworth: Penguin.

Ennis, G. (1997–98) *Preacher. Volume 5: Dixie Fried*. 1998, parts 27–33, New York: Vertigo/DC Comics.

Epigram (2001) 'Fanged Former Secretary', *Epigram*, 5 November, p. 5.

Fandrich, I. J. (1996) 'The Politics of Myth-Making: an analysis of the struggle for the "correct" appropriation of New Orleans voodoo queen Marie Laveau', *Social Compass*, **43(4)**, pp. 613–628.

Fanon, F. (1952) *Black Skin, White Masks*. 1986, London: Pluto Press.

Farrar, J. and Farrar, S. (1981) *The Witches' Bible: the complete witches' handbook*. Custer, WA: Phoenix Publishing Company.

Faucher, C. (2004) 'As the Wind Blows and Dews Came Down: ghost stories and collective memory in Singapore' in R. Bishop, J. Phillips and W.-W. Yeo (eds), *Beyond Description: Singapore, space, historicity*. London: Routledge.

Ferguson, H. (1996) *The Lure of Dreams: Sigmund Freud and the construction of modernity*. London: Routledge.

Flanders, S. (ed.) (1993) *The Dream Discourse Today*. London: Routledge in association with the Institute of Psycho-Analysis.

Fleurant, G. (1996) *Dancing Spirits: rhythms and rituals of Haitian Vodun, the Rada rite*. Westport, CT: Greenwood Press.

Florence, R. (1996) *Cities of the Dead: a journey through St Louis Cemetery #1, New Orleans, Louisiana*. Lafayette, LA: The Center for Louisiana Studies, University of Southwestern Louisiana.

Fong, J. and Fong, S. (2001) 'My Terrifying Journey to Hell: told by a "hungry ghost"!' in *There are Ghosts Everywhere in Singapore, Volume 3*. Singapore: The Publishing Consultant, pp. 43–61.

Foucault, M. (1963) *The Birth of the Clinic: an archaeology of medical perception*. 1973, London: Routledge.

Foucault, M. (1984) *The History of Sexuality. Volume 3: the care of the self*. 1988, Harmondsworth: Penguin.

Frayling, C. (1991) 'Lord Byron to Count Dracula' in C. Frayling (ed.), *Vampyres: Lord Byron to Count Dracula*. London: Faber & Faber, pp. 3–84.

French, P. (1972) *John Dee: the world of an Elizabethan Magus*. London: Routledge.

Freud, S. (1900) *The Interpretation of Dreams*. 1976, Harmondsworth: Volume 4, Penguin Freud Library.

Freud, S. (1901) *On Dreams*. Volume 5, *Standard Edition*. London: Hogarth Press.

Freud, S. (1905) 'Fragment of an Analysis of a Case of Hysteria: "Dora"' in *Case Histories I: 'Dora' and 'Little Hans'*. 1977, Harmondsworth: Volume 8, Penguin Freud Library, pp. 35–164.

Freud, S. (1907) 'Delusions and Dreams in Jensen's "Gradiva"' in *Art and Literature: Jensen's 'Gradiva', Leonardo Da Vinci and other works*. 1985, Harmondsworth: Volume 14, Penguin Freud Library, pp. 33–118.

Freud, S. (1908) 'Creative Writers and Day-Dreaming' in *Art and Literature: Jensen's 'Gradiva', Leonardo Da Vinci and other works*. 1985, Harmondsworth: Volume 14, Penguin Freud Library, pp. 131–141.

Freud, S. (1910) 'Five Lectures on Psycho-Analysis' in *Two Short Accounts of Psycho-Analysis*. 1962, Harmondsworth: Pelican Books, pp. 31–87.

Freud, S. (1911) 'The Handling of Dream-Interpretation in Psycho-Analysis'. Volume 12, *Standard Edition*. London: Hogarth Press, pp. 91–96.

Freud, S. (1913) 'Totem and Taboo' in *The Origins of Religion*. 1985, Harmondsworth: Volume 13, Penguin Freud Library, pp. 49–224.

Freud, S. (1914) 'On Narcissism: an introduction' in *On Metapsychology: the theory of psychoanalysis*. 1984, Harmondsworth: Volume 11, Penguin Freud Library, pp. 65–97.

Freud, S. (1915) 'The Unconscious' in *On Metapsychology: the theory of psychoanalysis*. 1984, Harmondsworth: Volume 11, Penguin Freud Library, pp. 167–222.

Freud, S. (1916–17) *Introductory Lectures on Psycho-Analysis*. Volumes 15–16, *Standard Edition*. London: Hogarth Press.

Freud, S. (1917a) 'A Metapsychological Supplement to the Theory of Dreams' in *On Metapsychology: the theory of psychoanalysis*. 1984, Harmondsworth: Volume 11, Penguin Freud Library, pp. 229–243.

Freud, S. (1917b) 'Mourning and Melancholia' in *On Metapsychology: the theory of psychoanalysis*. 1984, Harmondsworth: Volume 11, Penguin Freud Library, pp. 251–268.

Freud, S. (1919) 'The "Uncanny"' in *Art and Literature: Jensen's 'Gradiva', Leonardo Da Vinci and other works*. 1985, Harmondsworth: Volume 14, Penguin Freud Library, pp. 339–376.

Freud, S. (1920) 'Beyond the Pleasure Principle' in *On Metapsychology: the theory of psychoanalysis*. 1984, Harmondsworth: Volume 11, Penguin Freud Library, pp. 275–338.

Freud, S. (1921) 'Group Psychology and the Analysis of the Ego' in *Civilization, Society and Religion: group psychology, civilization and its discontents and other works*. 1985, Harmondsworth: Volume 12, Penguin Freud Library, pp. 95–178.

Freud, S. (1923) 'Remarks on the Theory and Practice of Dream-Interpretation'. Volume 19, *Standard Edition*, London: Hogarth Press.

Freud, S. (1925a) 'Some Additional Notes on Dream-Interpretation as a Whole'. Volume 19, *Standard Edition*. London: Hogarth Press.

Freud, S. (1925b) 'A Note upon the "Mystic Writing-Pad"' in *On Metapsychology: the theory of psychoanalysis*. 1984, Harmondsworth: Volume 11, Penguin Freud Library, pp. 429–434.

Freud, S. (1926) 'Inhibitions, Symptoms and Anxiety' in *On Psychopathology: inhibitions, symptoms and anxiety and other works*. 1979, Harmondsworth: Volume 10, Penguin Freud Library, pp. 237–333.

Freud, S. (1930) 'Civilization and its Discontents' in *Civilization, Society and Religion: group psychology, civilization and its discontents and other works*. 1985, Harmondsworth: Volume 12, Penguin Freud Library, pp. 251–340.

Freud, S. (1933) *New Introductory Lectures on Psychoanalysis*. 1973, Harmondsworth: Volume 2, Penguin Freud Library.

Freud, S. (1939) 'Moses and Monotheism: three essays' in *The Origins of Religion*. 1985, Harmondsworth: Volume 13, Penguin Freud Library, pp. 243–386.

Frisby, D. (2001) *Cityscapes of Modernity: critical explorations*. Cambridge: Polity Press.

Gaiman, N. (1988–96) *The Sandman*. 75 parts in 10 volumes, New York: DC Comics.

Gaiman, N. (1990–91) *The Books of Magic*. 4 parts in 1 volume, New York: Vertigo/DC Comics.

Gaiman, N. (1992) 'Soft Places' in N. Gaiman *The Sandman*. 1993, Volume VI, New York: DC Comics, pp. 124–148.

Gaiman, N. (1993a) 'Ramadan' in N. Gaiman, *The Sandman*. 1993, Volume VI, New York: DC Comics, pp. 226–258.

Gaiman, N. (1993b) 'A Tale of Two Cities' in N. Gaiman, *The Sandman*. 1994, Volume VIII, New York: DC Comics, pp. 18–41.

Gaiman, N. (1995) 'Hold Me' in N. Gaiman, *Midnight Days*. 1999, New York: DC Comics.

Gaiman, N. (1996) *Neverwhere*. London: Headline Book Publishing.

Garner, P. (2003) *A Walk Around Haunted London*. London: Louis' London Walks.

Gay, P. (1984) *The Bourgeois Experience: Victoria to Freud. Volume 1: The Education of the Senses*. New York: Oxford University Press.

Gehman, M. and Ries, N. (1988) *Women and New Orleans: a history*. New Orleans: Margaret Media, Inc.

Gelder, K. (1994) *Reading the Vampire*. London: Routledge.

Gelder, K. and Jacobs, J. M. (1998) *Uncanny Australia: sacredness and identity in a postcolonial nation*. Melbourne: Melbourne University Press.

Gelder, K. and Jacobs, J. M. (1999) 'The Postcolonial Ghost Story' in P. Buse and A. Stott (eds), *Ghosts: deconstruction, psychoanalysis, history*. Basingstoke: Macmillan, pp. 179–199.

Geschiere, P. (1998) 'Globalization and the Power of Indeterminate Meaning: witchcraft and spirit cults in Africa and East Asia', *Development and Change*, **29**, pp. 811–837.

Getty, S. (2002) 'Teen Vampire "Sliced Out Woman's Heart"', *Metro*, 17 July, p. 7.

Gilbert, E. (1998) '"Ornamenting the Façade of Hell": iconographies of 19th-century Canadian paper money', *Environment and Planning D: Society and Space*, **16(1)**, pp. 57–80.

Gilloch, G. (1996) *Myth and Metropolis: Walter Benjamin and the city*. Cambridge: Polity Press.

Gilman, S. (1993) *Freud, Race and Gender*. Princeton, NJ: Princeton University Press.

Glancey, J. (2002) 'The Dream Life of Buildings', the *Guardian*, 15 April, pp. 12–13.

Goh, P. K. (1997) *Origins of Chinese Festivals*. Singapore: Asiapac.

Goh, Y. K. and Wong, K. W. (2001) *The Sage and the Butterfly: a guide to Chinese dream interpretation*. Singapore: Rank Books.

Gordon, A. (1997) *Ghostly Matters: haunting and the sociological imagination*. Minneapolis, MN: University of Minnesota Press.

Hall, A. (2002) 'Vampires Get 28yrs', the Sun, 1 February, p. 27.

Hall, G. M. (1992) *Africans in Colonial Louisiana: the development of Afro-Creole culture in the eighteenth century*. Baton Rouge, LA: Louisiana State University Press.

Hannigan, J. (1998) *Fantasy City: pleasure and profit in the postmodern metropolis*. London: Routledge.

Haraway, D. (1995) 'Universal Donors in a Vampire Culture: it's all in the family, biological kinship categories in the twentieth-century United States' in W. Cronon (ed.), *Uncommon Ground: toward reinventing nature*. New York: W. W. Norton.

Harkness, D. (1999) *John Dee's Conversations with Angels: cabala, alchemy and the end of nature*. Cambridge: Cambridge University Press.

Harrison, S., Pile, S. and Thrift, N. (2004) 'Grounding Patterns: deciphering (dis)order in the entanglements of nature and culture' in S. Harrison, S. Pile and N. Thrift (eds), *Patterned Ground: the entanglements of nature and culture*. London: Reaktion Books, pp. 12–42.

Harvey, D. (1973) *Social Justice and the City*. London: Edward Arnold.
Harvey, D. (1988) 'Voodoo Cities', *New Statesman and Society*, 30 September, pp. 33–35.
Harvey, D. (1996) *Justice, Nature and the Geography of Difference*. Oxford: Basil Blackwell.
Harvey, D. (2000) *Spaces of Hope*. Edinburgh: Edinburgh University Press.
Hetherington, K. (2001) 'Phantasmagoria/Phantasm Agoria: materialities, spacialities, and ghosts', *Space and Culture*, special issue on Spatial Hauntings, **11/12**, pp. 24–41.
Hiller, S. (curator) (2000) *Dream Machines*. London: Hayward Gallery Publishing.
Hodgkin, K. and Radstone, S. (eds) (2003) *Contested Pasts: the politics of memory*. London: Routledge.
hooks, b. (2003) 'The Abyss of Surrender in Another Water' in U. Stahel (ed.), *If On A Winter's Night… Roni Horn…* . Winterthur/Göttingen: Winterthur Fotomuseum/Steidl Verlag.
Hooper, J. (2002) 'Blood-drinking Devil Worshippers Face Life for Ritual Satanic Killing', the *Guardian*, 1 February, p. 2.
Horn, R. (2000) *Another Water (The River Thames, For Example)*. Zurich: Scalo.
Howard, E. (1902) *Garden Cities of Tomorrow*. 1965, London: Faber & Faber.
Hurbon, L. (1993) *Voodoo: search for the spirit*. 1995, New York: Harry N. Abrams, Inc.
Imray, L. and Middleton, A. (1983) 'Public and Private: marking the boundaries' in E. Garmarnikow, D. Morgan, J. Purvis and D. Taylorson (eds), *The Public and the Private*. 1983, London: Heinemann, pp. 12–27.
Islington Gazette (2002) 'Vampires Shocker', *Islington Gazette*, 31 January, pp. 1–2.
Ivain, G. (1953) 'Formulary for a New Urbanism' in L. Andreotti and X. Costa (eds), *Theory of the Dérive and Other Situationist Writings on the City*. 1996, Barcelona: Museu d'Art Contemporani de Barcelona and ACTAR, pp. 14–17.
Jackson, R. (1981) *Fantasy: the literature of subversion*. London: Methuen.
Jacobs, J. (1961) *The Death and Life of Great American Cities*. Harmondsworth: Penguin.
Jacobs, J. M. (1996) *Edge of Empire: postcolonialism and the city*. London: Routledge.
Jaguaribe, B. (2001) 'Mestizo Memories: hybridity, eroticism and haunting', *Space and Culture*, special issue on Spatial Hauntings, **11/12**, pp. 87–105.
James, C. L. R. (1938) *The Black Jacobins: Toussaint L'Ouverture and the San Domingo revolution*. 1980, London: Allison and Busby.
Jameson, F. (1999) 'Marx's Purloined Letter' in M. Sprinker (ed.), *Ghostly Demarcations: a symposium on Jacques Derrida's Spectres of Marx*. London: Verso, pp. 26–67.
Johnson, J. (1991) *Congo Square in New Orleans*. 1995, New Orleans: Louisiana Landmarks Society.
Johnson, N. (1994) 'Sculpting Heroic Histories: celebrating the centenary of the 1798 rebellion in Ireland', *Transactions of the Institute of British Geographers*, **19**, pp. 78–93.
Johnson, N. (1995) 'Cast in Stone: monuments, geography and nationalism', *Environment and Planning D: Society and Space*, **13**, pp. 51–66.
Jones, E. (1931) 'On the Vampire' in C. Frayling (ed.), *Vampyres: Lord Byron to Count Dracula*. London: Faber & Faber, pp. 398–417.
Jones, R. (1999) *Walking Haunted London: 25 original walks exploring London's ghostly past*. London: New Holland Publishers.
Joyce, P. (2002) 'Maps, Blood and the City: the governance of the social in nineteenth-century Britain' in P. Joyce (ed.), *The Social in Question: new bearings in history and the social sciences*. London: Routledge, pp. 97–114.
Kaïka, M. and Swyngedouw, E. (2000) 'Fetishizing the Modern City: the phantasmagoria of urban technological networks', *International Journal of Urban and Regional Research*, **24(1)**, pp. 120–138.
Kaplan, L. (2003) 'Where the Paranoid Meets the Paranormal: speculations on spirit photography', *Art Journal*, **62(3)**, pp. 18–29.
Kennedy, M. (2003) 'Benjamin Franklin's House: the naked truth', the *Guardian*, 11 August, p. 24.
Kern, S. (1983) *The Culture of Space and Time 1880–1918*. Cambridge, MA: Harvard University Press.

Kerr, J. (2001) 'The Uncompleted Monument: London, war, and the architecture of remembrance' in I. Borden, J. Kerr, J. Rendell and A. Pivaro (eds), *The Unknown City: contesting architecture and social space*. Cambridge, MA: The MIT Press, pp. 68–89.

Khan, M. (1974) 'The Use and Abuse of Dream in Psychic Experience' in S. Flanders (ed.), *The Dream Discourse Today*. 1993, London: Routledge in association with the Institute of Psycho-Analysis, pp. 91–99.

Klein, V. C. (1996) *New Orleans Ghosts*. Metairie, LA: Lycanthrope Press.

Kósa, J. N. and Szablyár, P. (2002) *Underground Buda*. Budapest: City Hall.

Kristeva, J. (1980) *Powers of Horror: an essay on abjection*. 1982, New York: Columbia University Press.

Kuftinec, S. (1998) '[Walking through a] Ghost Town: cultural hauntologie in Mostar, Bosnia-Herzegovina or Mostar: a performance review', *Text and Performance Quarterly*, **18(2)**, pp. 81–95.

Laclau, E. (1995) 'The Time is Out of Joint' *Diacritics*, **25(2)**, pp. 86–96.

Ladd, B. (1997) *The Ghosts of Berlin: confronting German history in the Berlin landscape*. Chicago: University of Chicago Press.

Laplanche, J. and Pontalis, J. B. (1973) *The Language of Psychoanalysis*. 1988, London: The Institute of Psycho-Analysis/Karnac Books.

Lash, S. and Urry, J. (1994) *Economies of Signs and Space*. London: Sage.

Le Corbusier (1927) *Towards a New Architecture*. 1982, London: Butterworth.

Le Corbusier (1933) *The Radiant City*. 1967, New York: Orion Press.

Le Goff, J. (1985) *L'Imaginaire Médiéval*. Paris: Gallimard.

Lee, R. (1989) 'Bukit Batok' in R. Lee (ed.), *True Singapore Ghost Stories. Volume 1*. Singapore: Angsana Books, pp. 66–68.

Lee, R. (ed.) (1995a) *True Singapore Ghost Stories, Volume 6*. Singapore: Angsana Books.

Lee, R. (ed.) (1995b) *True Singapore Ghost Stories, Volume 7*. Singapore: Angsana Books.

Lees, L. (2004) 'The Emancipatory City: urban (re)visions' in L. Lees (ed.), *The Emancipatory City: paradoxes and possibilities*. London: Sage, pp. 3–20.

Lefebvre, H. (1974) *The Production of Space*. 1991, Oxford: Basil Blackwell.

Leyshon, A. and Thrift, N. (1997) *Money/Space: geographies of monetary transformation*. London: Routledge.

Lip, E. (1997) *What is Feng Shui?* London: Academy Editions.

Little, J., Peake, L. and Richardson, P. (eds) (1988) *Women in Cities: gender and the urban environment*. London: Macmillan Education.

Lovell, N. (2002) *Cord of Blood: possession and the making of Voodoo*. London: Pluto Press.

Mace, R. (1976) *Trafalgar Square: emblem of Empire*. London: Lawrence & Wishart.

Marcus, L. (ed.) (1999) *Sigmund Freud's The Interpretation of Dreams: new interdisciplinary essays*. Manchester: Manchester University Press.

Marcus, S. (1999) *Apartment Stories: city and home in nineteenth-century Paris and London*. Berkeley: University of California Press.

Marx, K. (1843) 'Letters from the Franco-German Yearbooks' in K. Marx, *Early Writings*. 1975, Harmondsworth: Penguin, pp. 199–209.

Marx, K. (1850) 'The Class Struggle in France, 1848–1850' in K. Marx, *Surveys from Exile*. 1973, Harmondsworth: Penguin, pp. 35–142.

Marx, K. (1852) 'The Eighteenth Brumaire of Louis Bonaparte' in K. Marx, *Surveys from Exile*. 1973, Harmondsworth: Penguin, pp. 143–249.

Marx, K. (1857–58) *Grundrisse: foundations of the critique of political economy (rough draft)*. 1973, Harmondsworth: Penguin.

Marx, K. (1867) *Capital, Volume 1*. 1976, Harmondsworth: Penguin.

Marx, K. and Engels, F. (1872) *The Communist Manifesto*. 1967, Harmondsworth: Penguin.

Maspero, F. (1990) *Roissy Express: a journey through the Paris suburbs*. 1994, London: Verso.

McDowell, L. (1997) *Capital Culture: gender at work in the city*. Oxford: Basil Blackwell.

McLean, H. and Cole, A. (2001) *The DreamCatchers Handbook: learn to understand the personal significance of your dreams*. London: Carlton Books.

Métraux, A. (1959) *Voodoo in Haiti*. 1972, New York: Schocken Books.

Mighall, R. (1999) *A Geography of Victorian Gothic Fiction: mapping history's nightmares*. Oxford: Oxford University Press.

Mitchell, T. (1999) 'Dreamland: the neoliberalism of your desires', *Middle East Report*. Spring, pp. 28–33.

Moench, D. (1991) *Batman and Dracula: red rain*. New York: DC Comics.

Moey, N. (1990) *Pontianak: 13 chilling tales*. Singapore: Times Books International.

Mooney, G., Pile, S. and Brook, C. (1999) 'On Orderings and the City' in S. Pile, C. Brook and G. Mooney (eds), *Unruly Cities? Order/Disorder*. London: Routledge in association with the Open University, pp. 345–363.

Moorcock, M. (2001) *London Bone*. London: Scribner.

Moore, A. (1985) 'Growth Patterns' in A. Moore, *The Swamp Thing: the curse*. 2000, New York: Vertigo/DC Comics, pp. 51–72.

Moore, A. (1999) *Voodoo: dancing in the dark*. 4 parts in 1 volume, New York: Wildstorm/DC Comics.

Moore, A. (1999–2000) *The League of Extraordinary Gentlemen*. 6 parts in 1 volume, La Jolla, CA: America's Best Comics.

Moore, T. (1997) *The Re-Enchantment of Everyday Life*. London: Harper Collins.

Moretti, F. (1978a) 'Dialectics of Fear' in F. Moretti, *Signs Taken for Wonders: essays in the sociology of literary forms*. 1988, London: Verso, pp. 83–108.

Moretti, F. (1978b) 'Clues' in F. Moretti, *Signs Taken for Wonders: essays in the sociology of literary forms*. 1988, London: Verso, pp. 130–156.

Morrison, G. (1998–99) *The Invisibles. Volume 2: Kissing Mister Quimper*. 2000, Parts 14–22, New York: DC Comics.

Morrison, T. (1987) *Beloved*. London: Picador.

Nadkarni, M. (2003) 'The Death of Socialism and the Afterlife of its Monuments: making and marketing the past in Budapest's Statue Park' in K. Hodgkin and S. Radstone (eds), *Contested Pasts: the politics of memory*. London: Routledge, pp. 193–207.

Nast, H. J. (2000) 'Mapping the "unconscious": racism and the Oedipal family', *Annals of the Association of American Geographers*, **90(2)**, pp. 215–255.

National Archives of Singapore (1996) *The Japanese Occupation 1942–1945: a pictorial record of Singapore during the War*. Singapore: Times Editions.

Neate, P. (2001) *Twelve Bar Blues*. Harmondsworth: Penguin.

Niehaus, I. (2001) 'Witchcraft in the New South Africa: from colonial superstition to postcolonial reality?' in H. Moore and T. Sanders (eds), *Magical Interpretations, Material Realities: modernity, witchcraft and the occult on postcolonial Africa*. London: Routledge, pp. 184–205.

Niehaus, I. Mohlala, E. and Shokane, K. (2001) *Witchcraft, Power and Politics: exploring the occult in the South African Louveld*. London: Pluto Press.

Noble, S. (1996) *Feng Shui in Singapore*. Singapore: HEIAN International Publishers.

Nora, P. (1989) 'Between Memory and History: les lieux de mémoire', *Representations*, **26** (Spring), pp. 7–25.

Ong, J. (2001) 'Festival of the Hungry Ghosts (15th Day of the Seventh Moon)' in *There are Ghosts Everywhere in Singapore, Volume 3*. Singapore: The Publishing Consultant, pp. 62–64.

Osborne, T. (1996) 'Security and Vitality: drains, liberalism and power in the nineteenth century' in A. Barry, N. Rose and T. Osborne (eds), *Foucault and Political Reason: liberalism, neo-liberalism and rationalities of government*. London: University College Press.

Park, R. E. (1925a) 'The City: suggestions for investigation of human behavior in the urban environment' in R. E. Park and E. W. Burgess with R. D. McKenzie and L. Wirth, *The City: suggestions for investigation of human behavior in the urban environment*. 1984, Chicago: Midway Reprint, University of Chicago Press, pp. 1–46.

Park, R. E. (1925b) 'Magic, Mentality, and City Life' in R. E. Park and E. W. Burgess with R. D. McKenzie and L. Wirth, *The City: suggestions for investigation of human behavior in the urban environment*. 1984, Chicago: Midway Reprint, University of Chicago Press, pp. 123–141.

Penczak, C. (2001) *City Magick: urban rituals, spells, and shamanism*. York Beach, ME: Weiser Books.

Pick, D. and Roper, L. (2000) 'Psychoanalysis, Dreams, History: an interview with Hanna Segal', *History Workshop Journal*, **49**, pp. 161–170.

Pile, S. (1991) 'The Un(known)City … or, an urban geography of what lies buried below the surface' in I. Borden, J. Kerr, A. Pivaro and J. Rendell (eds), *The Unknown City: contesting architecture and social space*. Cambridge, MA: The MIT Press, pp. 262–279.

Pile, S. (1996) *The Body and the City: psychoanalysis, subjectivity and space*. London: Routledge.

Pile, S. (1998) 'Freud, Dreams and Imaginative Geographies' in A. Elliott (ed.), *Freud 2000*. Cambridge: Polity Press, pp. 204–234.

Pile, S. (1999) 'What is a City?' in D. Massey, J. Allen and S. Pile (eds), *City Worlds*, London: Routledge, pp. 3–52.

Pile, S. (2000) 'Sleepwalking in the Modern City: Walter Benjamin, Sigmund Freud and the world of dreams' in S. Watson and G. Bridge (eds), *Blackwell Companion to Urban Studies*. Oxford: Basil Blackwell, pp. 75–86.

Pile, S. (2002) 'Memory and the City' in J. Campbell and J. Harbord (eds), *Temporalities: autobiography in a postmodern age*. Manchester: Manchester University Press, pp. 111–127.

Pile, S. (2003) 'Struggles over Geography' in K. Anderson, M. Domosh, S. Pile and N. Thrift (eds), *Handbook of Cultural Geography*. London: Sage, pp. 23–30.

Pile, S., Brook, C. and Mooney, G. (eds) (1999) *Unruly Cities? Order/Disorder*. London: Routledge in collaboration with the Open University.

Pile, S. and Thrift, N. (eds) (2000) *City A Z*. London: Routledge.

Pinder, D. (1996) 'Subverting Cartography: the situationists and maps of the city', *Environment and Planning A*, **28(3)**, pp. 405–427.

Plant, S. (1992) *The Most Radical Gesture: the Situationist International in a postmodern age*. London: Routledge.

Poe, E. A. (1839) 'The Man in the Crowd' in *Selected Tales*. 1998, Oxford: Oxford University Press, pp. 84–91.

Poe, E. A. (1844) 'The Purloined Letter' in *Selected Tales*. 1998, Oxford: Oxford University Press, pp. 249–265.

Polidori, J. (1819) 'The Vampyre' in C. Frayling (ed.), *Vampyres: Lord Byron to Count Dracula*. London: Faber & Faber, pp. 107–125.

Powell, K. (1992) *Vauxhall Cross: the story of the design and construction of a new London landmark*. London: Wordsearch Publishing.

Praoveanu, I. (1999) *Bran Castle*. Brasov, Romania: C2 Design House Publishing.

Prendergast, C. (1992) *Paris and the Nineteenth Century*. Oxford: Basil Blackwell.

Quinodoz, J.-M. (2002) *Dreams That Turn Over a Page*. London: Routledge.

Radcliffe, S. (1993) 'Women's Place/*El Lugar de Mujeres*: Latin American and the politics of gender identity' in M. Keith and S. Pile (eds), *Place and the Politics of Identity*. London: Routledge, pp. 102–116.

Rangan, H. (2003) 'The Muti Trade: South Africa's indigenous medicines', www.arts.monash.edu.au/ges/who/muti/muti_trade1.html. Last accessed 25 July 2004.

Rankin, I. (2001) 'Welcome to the Spook Capital of the World', the *Independent*, 22 April, p. 26.

Rendell, J. (2002) *The Pursuit of Pleasure: gender, space and architecture in Regency London*. London: Athlone Press.

Resnik, S. (1987) *The Theatre of the Dream*. London: Routledge.

Rice, A. (1976) *Interview with the Vampire*. London: Warner Books.

Rickels, L. A. (1999) *The Vampire Lectures*. Minneapolis, MN: University of Minnesota Press.

Roach, J. (1996) *Cities of the Dead: circum-atlantic performance*. New York: Columbia University Press.

Robinson, J. (1998) '(Im)mobilizing space-dreaming of change' in H. Judin and I. Vladislavic (eds), *Blank _____ : architecture, apartheid and after*. Rotterdam: Netherlands Architecture Institute, pp. 163–171.

Robinson, J. (2000) 'Feminism and the Spaces of Transformation', *Transactions of the Institute of British Geographers*, **25(3)**, pp. 285–301.

Robinson, J. (2002) 'Global and World Cities: a view from off the map', *International Journal of Urban and Regional Research*, **26(3)**, pp. 531–554.

Robinson, J. (forthcoming) *The Ordinary City: between modernity and development*. London: Routledge.

Royle, N. (2003) *The Uncanny*. Manchester: Manchester University Press.

Ruddick, S. (2004) 'Domesticating Monsters: cartographies of difference and the emancipatory city' in L. Lees (ed.), *The Emancipatory City: paradoxes and possibilities*. London: Sage, pp. 23–39.

Sack, R. D. (1980) *Conceptions of Space in Social Thought: a geographic perspective*. London: Macmillan.

Sadler, S. (1998) *The Situationist City*. Cambridge, MA: The MIT Press.

Saghiyeh, H. (2001) 'Al-Qaeda Loses Itself in Dream World', the *Observer*, 16 December, p. 5.

Sandercock, L. (ed.) (1998) *Making the Invisible Visible: a multicultural planning history*. Berkeley: University of California Press.

Sandercock, L. (2002) 'Difference, Fear and Habitus: a political economy of urban fears' in J. Hillier and E. Rooksby (eds), *Habitus: a sense of place*. Aldershot: Ashgate Press, pp. 203–218.

Sanders, J. (2002) *Celluloid Skyline: New York and the movies*. London: Bloomsbury.

Sassen, S. (2001) *The Global City: New York, London, Tokyo* (Second Edition). Princeton, NJ: Princeton University Press.

Schivelbusch, W. (1983) *Disenchanted Night: the industrialization of light in the nineteenth century*. 1995, Berkeley: California University Press.

Schlör, J. (1998) *Nights in the Big City: Paris, Berlin, London 1840–1930*. London: Reaktion Books.

Schnitzler, A. (1926) *Dream Story*. Harmondsworth: Penguin.

Schoonover, K. (2003) 'Ectoplasms, Evanescence, and Photography', *Art Journal*, **62(3)**, pp. 30–43.

Schopp, A. (1997) 'Cruising the Alternatives: homoeroticism and the contemporary vampire', *Journal of Popular Culture*, **30(4)**, pp. 231–244.

Scott, K. (2001) 'Researchers Enter into the Spirit of Edinburgh's Underground Ghosts', the *Guardian*, 17 April, front page.

Segal, H. (1990) *Dream, Phantasy and Art*. London: Routledge.

Sennett, R. (1970) *The Uses of Disorder: personal identity and city life*. 1996, London: Faber & Faber.

Sennett, R. (1994) *The Flesh and the Stone: the body and the city in western civilisation*. London: Faber & Faber.

Sharpe, J. (2003) *Ghosts of Slavery: a literary archaeology of black women's lives*. Minneapolis, MN: University of Minnesota Press.

Sheringham, M. (ed.) (1996) *Parisian Fields*. London: Reaktion Books.

Sibley, D. (1995) *Geographies of Exclusion*. London: Routledge.

Simmel, G. (1900) *The Philosophy of Money*. 1990, London: Routledge.

Simmel, G. (1903) 'The Metropolis and Mental Life' in P. Kasinitz (ed.), *Metropolis: centre and symbol of our times*. 1995, Basingstoke: Macmillan, pp. 30–45.

Sinclair, I. (1975) *Lud Heat: a book of dead hamlets May 1974 to April 1975*. 1995, London: Granta Books.

Sinclair, I. (1991) *Downriver (Or, The Vessels of Wrath): a narrative in twelve tales*. London: Paladin Books.

Sinclair, I. (1997) *Lights Out for the Territory: 9 excursions in the secret history of London*. London: Granta Books.

Slater, D. (2003) 'Beyond Euro-Americanism – democracy and post-colonialism' in K. Anderson, M. Domosh, S. Pile and N. Thrift (eds), *Handbook of Cultural Geography*. London: Sage, pp. 420–432.

Smith, K. (1998) *Journey into Darkness... Ghosts and Vampires of New Orleans*. New Orleans: De Simonin Publications.

Smith, P. (2002) 'The Witchdoctor who Trades "Slaughtered Children's Fingers" as Voodoo Sex Charms', *Daily Express*, 3 November, pp. 18–19.

Sobel, M. (2000) *Teach Me Dreams: the search for self in the revolutionary era*. Princeton, NJ: Princeton University Press.

Soja, E. (1996) *Thirdspace: journeys to Los Angeles and other real-and-imagined places*. Oxford: Basil Blackwell.

Soja, E. (2001) 'On Spuitstraat: the contested streetscape of Amsterdam' in I. Borden, J. Kerr, J. Rendell and A. Pivaro (eds), *The Unknown City: contesting architecture and social space*. Cambridge, MA: The MIT Press, pp. 280–295.

Sorkin, M. (2003) *Starting from Zero: reconstructing downtown New York*. New York: Routledge.

Spain, D. (2001) *How Women Saved the City*. Minneapolis, MN: University of Minnesota Press.

St Clair, J. (1999) 'Seattle Diary: it's a gas, gas, gas', *New Left Review*, **238**, pp. 81–96.

Stallybrass, P. and White, A. (1986) *The Politics and Poetics of Transgression*. London: Methuen.

States, B. (1997) *Seeing in the Dark: reflections on dreams and dreaming*. New Haven, CT, and London: Yale University Press.

Stephen, M. (1998) 'Consuming the Dead: a Kleinian perspective on death rituals cross-culturally', *International Journal of Psychoanalysis*, **79**, pp. 1173–1194.

Stephen, M. (1999) 'Witchcraft, Grief, and the Ambivalence of Emotions', *American Ethnologist*, **26(3)**, pp. 711–737.

Stoker, B. (1897) *Dracula*. 1994, Harmondsworth: Penguin.

Tallant, R. (1946) *Voodoo in New Orleans*. 1998, Gretna: Pelican.

Taussig, M. (1997) *The Magic of the State*. London: Routledge.

Taylor, P. (2003) *World City Network: a global urban analysis*. London: Routledge.

Tedlock, B. (ed.) (1992) *Dreaming: anthropological and psychological interpretations*. Santa Fe, NM: School of American Research Press.

Tester, K. (ed.) (1994) *The Flâneur*. London: Routledge.

Thomson, J. (1880) *The City of Dreadful Night*. 1998, Edinburgh: Canongate Books.

Till, K. (1999) 'Staging the Past: landscape designs, cultural identity and *Erinnerungspolitik* at Berlin's Neue Wache', *Ecumene*, **6(3)**, pp. 251–283.

Till, K. (2005) *The New Berlin: memory, politics, place*. Minneapolis, MN: University of Minnesota Press.

Titmuss, R. M. (1997) *The Gift Relationship: from human blood to social policy* (Updated and Expanded Edition). New York: New Press.

Too, L. (2001) *Irresistible Book of Feng Shui Magic: 48 sure ways to create magic in your living space*. London: Element Publishers.

Trigo, B. (1999) 'Anemia and Vampires: figures to govern the colony, Puerto Rico, 1880–1904', *Comparative Studies in Society and History*, **41(1)**, pp. 104–123.

Trouillot, M.-R. (1995) *Silencing the Past*. Boston: Beacon.

True Ghost Stories (2000a) *Pontianak Nightmares*. Singapore: ASURAS.

True Ghost Stories (2000b) *Vampires Nightmares*. Singapore: ASURAS.

Tuan, Y.-F. (1978) 'The City: its distance from nature', *Geographical Review*, **68**, pp. 1–12.

Vasager, J. (2002) 'Muti: goal of human sacrifice', the *Guardian*, 20 April, p. 8.

Vidler, A. (1992) *The Architectural Uncanny: essays in the modern unhomely*. Cambridge, MA: The MIT Press.

Vidler, A. (2000) *Warped Space: art, architecture, and anxiety in modern culture*. Cambridge, MA: The MIT Press.

Walkowitz, J. R. (1992) *City of Dreadful Delight: narratives of sexual danger in Late-Victorian London.* London: Virago.

Ward, G. (2000) *Cities of God.* London: Routledge.

Ware, V. and Back, L. (2002) *Out of Whiteness: color, politics, and culture.* Chicago: University of Chicago Press.

Watts, M. J. (1999) 'Collective Wish Images: geographical imaginaries and the crisis of national development' in D. Massey, J. Allen and P. Sarre (eds), *Human Geography Today.* London: Polity Press, pp. 85–107.

Weigel, S. (1996) *Body- and Image-Space: re-reading Walter Benjamin.* London: Routledge.

White, L. (1995) '"They Could Make Their Victims Dull": genders and genres, fantasies and cures in colonial Southern Uganda', *American Historical Review*, **100(5)**, pp. 1379–1402.

White, L. (2000) *Speaking with Vampires: rumor and history in colonial history.* Berkeley: University of California Press.

Williams, R. (1973) *The Country and the City.* 1985, London: Hogarth Press.

Williams, R. (1989) *Resources of Hope.* London: Verso.

Williamson, M. (2001) 'Vampires and the Gendered Body' in N. Watson and S. Cunningham-Burley (eds), *Reframing the Body.* Basingstoke: Palgrave, pp. 96–112.

Wilson, E. (1991) *The Sphinx in the City: urban life, the control of disorder, and women.* London: Virago.

Wilson, M. O. (1998) 'Dancing in the Dark: the inscription of blackness in Le Corbusier's Radiant City' in H. Nast and S. Pile (eds), *Places through the Body.* London: Routledge, pp. 133–152.

Wirth, L. (1938) 'Urbanism as a Way of Life', *American Journal of Sociology*, **44**, pp. 1–24.

Woods, C. (1998) *Development Arrested: the blues and plantation power in the Mississippi delta.* London: Verso.

Woolf, V. (1932) 'The London Scene: abbeys and cathedrals' in V. Woolf, *The Crowded Dance of Modern Life. Selected Essays: Volume 2.* 1993, Harmondsworth: Penguin, pp. 122–127.

Woolley, B. (2002) *The Queen's Conjuror: the life and magic of Dr Dee.* London: Flamingo.

Wright, P. (2001) 'Around the World in Three Hundred Yards' in I. Borden, J. Kerr, J. Rendell and A. Pivaro (eds), *The Unknown City: contesting architecture and social space.* Cambridge, MA: The MIT Press, pp. 492–503.

Young, I. M. (1990) *Justice and the Politics of Difference.* Princeton, NJ: Princeton University Press.

Young, R. (1995) *Colonial Desire: hybridity in theory, culture and race.* London: Routledge.

Zizek, S. (1989) *The Sublime Object of Ideology.* London: Verso.

Index

'A Berlin Chronicle' 137
Abney Park Cemetery 7
advertisements, dreams 27, 31, 32
agglomeration 18
Albert Embankment *see* Vauxhall Cross
Alexandra Hospital 132
All Saints Day 82–3
Alourdes 71, 72
'A Lovely Dream' 43–5, 47
ambivalence
   blood 97
   magic 65
   money 87
angels 23–4, 175
Angelus Novus 139, 181–2
anti-capitalism protests 52, 136,
     156–62, 178–9
anti-colonialism 151–6
anxieties
   dreams 44, 45
   space-work 49
Arcades Project 50, 56
Augé, Marc 30

Baedeker, 106
Baltimore 176–7, 180
Báthory, Countess Elizabeth 102, 126
Baudelaire, C. 20, 104–5, 115
Bellamy, Edward 176
Bell, Michael E. 103
Bell, Michael M. 135–6
Benjamin, Walter 19–21, 25, 27–8,
     49–58, 165–6, 181–2
   awakening 53–6, 175
   Baudelaire 104–5

Benjamin, Walter *cont.*
   childhood 136, 137–9
   context 167
   ghosts 160
   magic 59–60, 61, 95
Berlin 17–18, 136, 137–9
Bielz, E. A. 106
Bienville 126
Blair, Tony 158
blood 96, 97–9, 120
   race 122
   see also vampires
blood-work 104, 125, 127–8, 169, 174
Boner, Charles 106
boredom 13
Boutique du Vampyre 124
Bringing out the Dead 149–51
Brite, Poppy 123–4, 125
British Museum 68
Brolin, R. 77, 78–9
Brown, K. M. 71–2
Budapest 23–4
Buffy the Vampire Slayer 100, 128
buildings
   feng shui 88–9, 90–1
   magic 68–9
Bukit Batok 146–7

Calmet, Dom Augustin 103
Calvino, Italo 25–6, 34–5
Canary Wharf 5, 7, 68, 90, 153, 154
capitalism
   anti-capitalism protests 52, 136,
     156–62, 178–9
   dreams 30, 31

Carayol, R. 86–7
Caribbean Islands 62–3, 65–6
Castle, Terry 19
Catholicism, Voodoo links 71, 73, 81
causality, dream-work 48
Cenotaph 159–60, 161
censorship 46
Chamani, Priest Oswan 81
Chesterton, G. K. 50
Chingford Mount 5–6
Christaller, 179
City of Angels 100
city-work 48, 51
civic toilette 110
Claiborne, William 76
Class War 5
Code Noir 76
commodities 55
commodity capitalism 30, 31
condensation 41, 42, 44–5, 46
Congo Square 73–4, 75–8, 82, 83, 95, 118, 175
Connelly, Joe 149–51
convergence 44
Copjec, J. 98, 116
Corbin, Alain 110
Corporate Voodoo 86–7
cosmos 69, 71
Crary, J. 19
Creoles 120, 143–4
crossroads 83

dance, Congo Square 77–8
Daudet, Alphonse 44
Debord, Guy 4, 12–15, 19
dérive 12, 13
Derrida, Jacques 142–3, 156, 160
desire
    dreams 34, 35, 44, 55–6, 167
    money 86
    space-work 49
    vampires 98, 112–17
détournment 13
dialectical images 56, 163
diminution 45
displacement 41, 42, 45–6, 57
divergence 45
Donald, James 2

Dracula 101, 102, 106–9, 111, 112, 115, 116–17, 175
Dreamachine 42
dreams 3, 20–2, 24, 25–58, 138, 165–8, 175–81
    analysis 41, 43–4, 166, 167, 181
    awakening 53–6, 175
    cities of imagination 35–41
    city space 47–53
    elements 43–4
    factory 2
    meaning 30, 41
    spaces 47
    space-work 172–3
    symbols 43–4, 45, 46, 48
    thoughts 44, 45, 47
    web 51
Dream Story 34
dream-work 21, 27, 40–9, 56–8, 166, 167–8, 171–2
drifting 12, 13

earth 69
economics
    politics 171
    Voodoo 84–7
Ellis, B. 105
Ellis, Warren 148
emotional life 4, 18
emotional-work 3, 48, 166–8, 171–2
Engels, F. 21, 109–10
etiquettes 18
Eudelle, Rose 65
exaggeration 45
exchange, Voodoo 85–6

Fanon, Frantz 93
fantasy 34
Faraday Centre 91, 92
Faucher, Carole 113
feng shui 61, 87–91, 95
Festival of the Hungry Ghosts 133–4
Firth, D. 86–7
Franklin, Benjamin 147
Frayling, Christopher 102, 103, 117
free association 43
Freud, Sigmund 61–2
    Benjamin's work 50, 55

Freud, Sigmund *cont.*
  blood 98
  dreams 21, 26, 27, 28, 40–8, 57–8, 166, 167, 175
  ghosts 134
  grief-work 149
  in/external worlds 69
  magic 63–5, 95
  offerings 80–1
  the uncanny 39–40, 140–1, 171
  unconscious 39, 41

Gaiman, Neil 2, 27, 35–40, 50, 145
Gandolfo, C. M. 64
Gerard, Dorothea 106
Gerard, Emily 106
ghosts 3, 14–15, 21, 24, 129–64, 166, 170–1
  anti-capitalism protests 156–62
  Berlin 136, 137–9
  London 6, 8, 136, 144–5, 147–8, 151–62
  New Orleans 136, 143–4, 147–8
  occult spatialities 173, 174–5
  rats 23
  Singapore 132–5, 136, 145–8
globalisation 173–4
  magic 66, 91–4
  occult 94
  undead 128–9
Gordon, Avery 131, 132, 151, 171
graffiti 5, 6, 13, 69
Grando, Giure 102
grief-work 149, 155, 163–4, 170–2
guerrilla gardening 157, 158
Gysin, Brion 42

Hackert, Frank 99–100
Hall, Allan 99
Hall, Gwendolyn 74
Halloween 81, 82, 124
Hamlet 132, 142–3
Harvey, David 84, 176–80
Heinle, Fritz 138
Hellblazer 36, 148
Heroes Square, Budapest 23
Hooper, John 99–100
Horn, Roni 145
Howard, Ebenezer 176, 177

Howard, Michael 32–3
Hui, Ann 133

Imagining the Modern City 2
inhibitions, dreams 44
injustice 55
intellectualism 17
Interview with the Vampire 101, 118–22
Invisible Cities 25–6
IRA 11, 152–5
Ivain, Gilles 13–15

Jackson Square 80, 81, 82, 83, 84, 125
Jacobs, Jane 142
Jameson, Fredric 160
Johnson, J. 76–7, 78
Jones, Ernest 98, 102, 103
Jordan, Neil 118, 120
Joyce, Patrick 110

Keiller, Patrick 4–12, 13, 14, 19, 56
key nodes 68
King, Martin Luther 33
Klee, Paul 139, 181–2
Knight, Vlad Tepes 125
Kray, Ronnie 6
Kublai Khan 25–6

Lafayette Cemetery 122
LaLaurie, Delphine Macarty 126
LaLaurie House 125, 126
LaLaurie, Nicholas 126
land 69
latent content 42
Latrobe, Benjamin 77
Laveau, Marie 73, 77–80
Le Corbusier 177
Lee, R. 146
Lefebvre, H. 35, 181
Legba, Papa 83
Libeskind, Daniel 161
Lion, Ferdinand 51
Lip, Evelyn 88
Lola, Mama 71, 72
London 175, 177
  anti-capitalism protests 52, 136, 156–62, 178–9
  ghosts 6, 8, 136, 144–5, 147–8, 151–62

London *cont.*
  magic 68, 91–4
  psychogeography 4–12
  vampires 99–100, 105, 107–9
Lone Woman 113–14
Lost Souls 123–4
Lund Cathedral 22–3

McLean, Helen 31–2
magic 3, 21, 24, 58, 59–96, 166, 168–9, 171
  ambivalence 65
  economy 84–7
  London 68, 91–4
  New Orleans 68, 73–84
  New York 66–72
  science 62, 63, 65
  Singapore 87–91
  spatiality 64
  see also Voodoo
magic-work 64, 69, 168–9
Mai Mai market 91
Manchester, Sean 105
Mandela, Nelson 93
mapping cities 137
Marcus, Sharon 147
Marx, Karl 53, 105, 171
  dreams 25
  ghosts 131, 132, 156
  modernity 21
  money 111–12
The Matrix 33–4
melancholia 149, 155–6, 163
memories 137–9
  involuntary 136
mentality, magic 61–2, 63
mental states 15–19
MI6 10, 152–3, 154
Mighall, R. 116
Miriam, Priestess 81–3
misers 85
money 17, 18
  dreams 31
  magic 85–6, 87
  vampires 111–12
Moore, Alan 36
morals, dreams 34
Moretti, F. 31, 116
murder, 'Adam' 61, 91, 92–4

muti trade 91, 92, 93
Mutwa, Credo 93, 94

New Orleans 175
  ghosts 136, 143–4, 147–8
  magic 68, 73–84
  vampires 100, 101, 117–27
  Voodoo Spiritual Temple 81–2, 84
New York 175
  ghosts 148–51
  magic 66–72
  World Trade Center 161

obeah 62, 65–6, 81
occult 59, 60, 95, 168
  globalisation 94
  rituals 105
  spatialities 172–5
  Voodoo dolls 64
'One-Way Street' 53–4, 56
Onogigovie, Sam 94
Orchard Road 114

Paole, Arnold 102
parallels 45
Paris 20, 147
  dreams 50
  vampires 104–5, 115, 121, 122
Park, Robert 1, 60–4, 65–6, 71, 95, 169
Parliament Square 156–7
Penczak, Chris 67–9, 72
personality 2–3, 30, 38
The Philosophy of Money 15–16
phobias, dreams 44
Piccadilly Circus 16
Plogojowitz, Peter 102, 103
Polidori, John 102
politics 178–9
  anti-capitalism protests 52, 136, 156–62, 178–9
  dreams 32–3
  economy 171
  ghosts 136
  science 63
Polo, Marco 25–6
pontianaks 101, 113–17, 128
poverty 55, 71, 72, 110
power relations 13, 48

proximity, magic 64
psychogeography 4–15

quadroons 143–4

Rangan, Haripriya 92
rationalism 17
representability issues 46, 57
representation 41, 42
repression
    dreams 49
    uncanny 140–1
Resources of Hope 176
reversal into opposites 45
rice 74
Rice, Anne 101, 118–22, 125
Rickels, Laurence 102, 103
ritual murder 61, 91, 92–4
Robillard, David 46
Rose, Ruth Ellen 103
Royal Street 123, 125, 126, 143–4
Ruda, Daniel 99–100
Ruda, Manuela 99–100

St Louis Cathedral 80, 82, 83
St Louis Cemetery 118, 123, 125
The Sandman 35–40
sangoma 93
Schnitzler, Arthur 34
science, and magic 62, 63, 65
science fiction, dreams 33–4
secondary revision 41, 42, 43, 46–7, 57
sexuality, vampires 105, 112–17, 129
sexual liberation 34
Shakespeare, William 132, 142–3
shamanic rituals 69
sigil 69, 70
Siloso Beach 133
similarity, magic 64
Simmel, G. 4, 15–19, 40, 87, 95, 171
    indifference 155
    intellectual culture 180
    misers 85
    modernity 55
    money 111
Sinclair, Iain 4–12, 13, 14–15, 19, 22
Singapore 175
    feng shui 88–91

Singapore *cont.*
    ghosts 132–5, 136, 145–8
    magic 87–91
    vampires 112–17
Situationist International 4, 12–15
The Sixth Sense 136, 140–3, 151
sky 69
skyscapers 69
slavery 72, 74, 75–6, 78–9, 86, 144, 147
Smithfield Market 110, 148
Smith, K. 126
Smith, Paul 93
Smith, Sidney 125
Sobel, Mechal 30
social discourse, dreams 26
Sommerville, Ian 42
Sook Ching 133
sorcery 69–71
Southpark 29–30
Spaces of Hope 176–80
space-time 17–18, 64
space-work 48–9, 52, 89–90, 166, 172–5
spatiality
    dreams 41
    dream-work 47–9
    magic 64
    occult 172–5
spatiotemporal utopianism 178
Speaking with Vampires 103–4
Stallybrass, P. 110
Stoker, Bram 101, 102, 106–9, 112, 115, 116–17
structures of feelings 3
subsumption 45
Suntec City, 'The Fountain of Wealth' 88, 89
syncretic cities 73–84

taboo 63
Taoism 89–90, 133
telephone boxes 8, 9
terrorism 151–6
    IRA 11, 152–5
    Wandsworth Common bomb 11
Thomson, J. 177
thresholds
    ghosts 139
    magic 59–60, 67

thresholds *cont.*
  spaces 174–5
  vampires 170
time-space 17–18, 64
Times Square 66
time-work 48
Too, Lillian 89, 90
Torture Garden 100
Trafalgar Square 51–2, 156, 157–8, 179
Tranchepain, M. 126
transformations, dreams 28
transformative recognition 132
transvaluation, dreams 45
travel
  dreams 49
  psychogeography 4–12
turning 13

the uncanny 2, 39–40, 140–1, 171
unconscious 39, 41, 166
urban imaginaries 2
urbanism 162
Ursuline Convent 125, 126–7
Utopia 176, 177–9, 180

vampires 3, 21, 24, 96–130, 166, 169–70, 171
  London 99–100, 105, 107–9
  New Orleans 100, 101, 117–27
  occult spatialities 173–4, 175
  plagues 101, 102–6
  pontianaks 101, 113–17, 128
  Singapore 112–17
  see also Dracula
Vampire Tours 123–7
Vasager, J. 91
Vauxhall Cross 10, 151–3, 154–5
Vienna 34
Visible Secrets 133
Vodou 70–2, 95, 188

Voodoo Spiritual Temple 81–2, 84
Voodoo 61, 71–3, 95, 174, 175
  Atlantic 73–84
  dolls 64, 168
  economics 84–7
  London 93–4

Wandsworth Common bomb 11
The War of Dreams 30
watches 17–18
werewolves 22, 23, 175
West Indian island culture 62–3, 65–6
White, A. 110
White, Luise 103–4
Wilkinson, William 106
Williams, Raymond 3, 176
Wirth, Louis 100, 162
Woolf, Virginia 161, 165
work
  blood-work 104, 125, 127–8, 169, 174
  city-work 48, 51
  dream-work 21, 27, 40–9, 56–8, 166, 167–8, 171–2
  emotional-work 3, 48, 166–8, 171–2
  grief-work 149, 155, 163–4, 170–2
  magic-work 64, 69, 168–9
  space-work 48–9, 52, 89–90, 166, 172–5
  time-work 48
World Exhibition (1867) 50
World Trade Center 161

Yamashita, Lt General 132
yin and *yang* 89
Young, Hugo 158

zombie corporate culture 87

Indexed by Caroline Eley